Computing Applications in Molecular Spectroscopy

Computing Applications in Molecular Spectroscopy

Edited by
W. O. George
University of Glamorgan, Pontypridd, UK

D. Steele
Royal Holloway, University of London, Egham, UK

THE ROYAL SOCIETY OF CHEMISTRY

ISBN 0-85404-519-8

A catalogue record of this book is available from the British Library.

Published by The Royal Society of Chemistry,
Thomas Graham House, The Science Park, Cambridge
CB4 4WF
Typeset by Vision Typesetting, Manchester
Printed by Hartnolls Ltd, Bodmin, Cornwall, UK

Preface

The penetration of computing methods (or the more fashionable term – information technology) into many areas of work and every-day living has been a dominant feature of the second half of the twentieth century. Arguably so pervasive is this movement that computing should cease to be seen as a separate discipline or speciality but rather as a normal feature of the modern world.

It follows that a computer has no *raison d'être* unless it can be given a problem to solve or a purpose to serve. Some problems may be trivial to which the computer becomes the proverbial sledge-hammer. Other problems would otherwise be insoluble without powerful computers (the moon landing was a spectacular if symbolic example).

Why should spectroscopy merit special treatment in relation to computer opportunities? Any spectroscopic measurement requires a unification of the art, the science and the technology of the experimentalist. Until recent times the art and technology required photography or photoelectric-type recording of spectroscopic information. The volume of information is determined by the wavelength (frequency) range, resolution and the precision with which the intensity (absorbance) data are measured. This is potentially vast and therefore highly suited to computer storage. Depending on these conditions a single spectrum may occupy several kilobytes or several megabytes of computer memory.

In addition to storing information a computer, when on-line to a spectrometer, can reduce the time to collect and display a spectrum from several days to milliseconds. This is the technology that provides the spectral information which is the basis of the science of spectroscopy. But scrutiny of many of the spectra in this book, particularly those for samples in the vapour state, reveal them to be works of art. To many they inspire the same emotions as great paintings or great music. The underlying theories are also of great elegance and beauty in terms of the initial simplicity of concepts to which refinement is added to predict and explain details revealed by the more searching experiments.

Spectroscopic measurements can be converted into analytical, atomic or molecular information. The first has importance to industry and society; the last two have importance to scientific understanding of structure of matter. The science of spectroscopy has universal significance as demonstrated by the growth of astronomical measurements. Information on terrestrial and celestial samples or sources is interpreted from common scientific concepts and theory made accessible by modern computing power.

The community of interest in interfacing computing and spectroscopy spans industry, government and academia and is international. For this reason the contributors to this book are drawn from these three constituencies and from five European nations.

This book is an update and successor to an earlier volume* based on a series of six residential courses held at the University of Glamorgan and organized in conjunction with the Royal Society of Chemistry and the Infrared and Raman Discussion Group.

These previous ventures were strongly associated with Harry Willis who died suddenly, shortly after the earlier volume was published. He was the epitomy of the community wide cooperation, intellectual excitement and enquiry generated by this subject area. For these reasons this book is dedicated to the late Harry Willis and his family.

W. O. George
D. Steele

* 'Computer Methods in UV, Visible and IR Spectroscopy', ed. W. O. George and H. A. Willis, The Royal Society of Chemistry, Cambridge, 1990.

Contents

Contributors

A. N. Davies, *ISAS, Institut für Spektrochimie, Postfach 10.13.52 Bunsen Kirchhoff-Strasse 11, 4600 Dortmund 1, Germany*

J. R. Dixon, *University of Glamorgan, Pontypridd, Mid Glamorgan CF37 1DL, UK*

G. Downey, *TEAGASC, The National Food Centre, Dunsinea, Castleknock, Dublin 15, Eire*

M. A. Ford, *Perkin Elmer Limited, Beaconsfield, Buckinghamshire HP9 1QA, UK*

D. J. Gardiner, *University of Northumbria at Newcastle, Ellison Buildings, Ellison Place, Newcastle Upon Tyne NE1 8ST, UK*

W. O. George, *University of Glamorgan, Pontypridd, Mid Glamorgan CF37 1DL, UK*

A. S. Gilbert, *The Wellcome Research Laboratories, Langley Court, South Eden Park Road, Beckenham, Kent BR3 3BS, UK*

Rh. Lewis, *University of Glamorgan, Pontypridd, Mid Glamorgan CF37 1DL, UK*

H. J. Luinge, *University of Utrecht, Croesestraat 77a, 3522AD Utrecht, The Netherlands*

P. S. McIntyre, *University of Glamorgan, Pontypridd, Mid Glamorgan CF37 1DL, UK*

S. F. Parker, *Rutherford Appleton Laboratory, Chilton, Didcot, Oxfordshire OX11 0QX, UK*

D. Steele, *Royal Holloway, University of London, Egham, Surrey TW20 0EX, UK*

B. J. van der Veken, *Universitair Centrum Antwerpen, Groenenborgerlaan 171, B2020 Antwerpen, Belgium*

CHAPTER 1

General Data Handling

M. A. FORD

The whole field of computer data handling, which can be summarized in four brief statements, enables us to get from our data:

- What we want.
- Where we want.
- When we want.
- How we want.

  with the rider 'reliably'.

'What' covers the whole area of data processing for such things as qualitative and quantitative analysis – but many more also. 'Where' covers the field of data transmission and networking. 'When' is the subject of short and long term data storage and retrieval – archiving. 'How' relates to the way in which the results are to be presented? Tables? Graphical spectra? Contour plots? Quality report forms? Finally, reliability is essential if we are to have confidence in our results.

Major aspects of the 'what' and 'when' are covered elsewhere in this book; the purpose of this chapter is to give a broad introduction and to point out some issues and potential pitfalls which may otherwise be overlooked. 'How' is not covered elsewhere and the subject is expanded on a little at the end of the chapter. The reliability issue is covered more fully later.

1 GENERATION OF RAW DATA

Before any computational operations can take place it is necessary to convert the original spectroscopic data, which is almost always in analogue form (exceptions are 'counts' as in mass spectroscopy and photon counting) into digital format. To perform this digitization *without loss of information* certain rules must be followed. These rules were set out[1] in Chapter 1 of 'Computer Methods in UV, Visible and IR Spectroscopy' as were illustrations of the fact that original data are never truly 'raw' but are invariably subject to some arbitrary constraints. The reader is referred to the earlier work for details of the latter matter, but it is felt worthwhile

[1] 'Computer Methods in UV, Visible and IR Spectroscopy', ed. W. O. George and H. A. Willis, Royal Society of Chemistry, Cambridge, 1990.

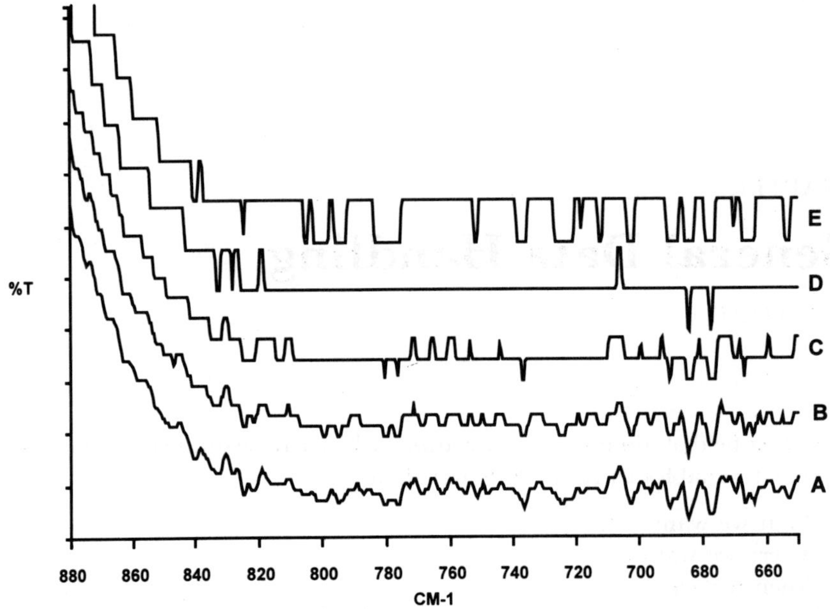

Figure 1 *Illustration of the effect of different ordinate data intervals on the digitization of spectral data. Data intervals: A, $\frac{1}{2}\times$ r.m.s. noise; B, $1\times$ r.m.s. noise; C, $2\times$ r.m.s. noise; D, $4\times$ r.m.s. noise; E as D with zero offset.*

repeating and explaining the rules for digitization here even though they are normally under the control of the instrument manufacturer, not the user.

1 The smallest measurable increment must be less than the noise.
2 The sampling rate must give at least two samples per cycle of the highest noise frequency.
3 The dynamic range must be sufficient to accept the largest signal.

Note: These rules are for spectral data. They are, of course, irrelevant to the digitization of, for example, molecular structures for which standard systems have been created.

Rule (1) ensures the correct averaging of noise. The factor by which the smallest step must be less than the noise cannot be defined precisely but it is generally accepted that a factor of 0.5 of the root mean square (RMS) noise value is satisfactory. The effect of non-observance of the rule is illustrated in Figure 1. Figure 1(A) shows the original data which was digitized with an increment of half the r.m.s. noise and Figure 1 (B), (C), (D) and (E) are replots after digitization with increments of (B) equal to, (C) twice, and (D) and (E) four times the RMS noise. In (B) the 'stepiness' is starting to be apparent while in (C), (D) and (E) it is severe. The difference between (D) and (E) is the purely chance situation of where the digitization step level falls in relation to the mean signal. Regrettably examples showing this 'digitization noise' are all too common.

One further important point to note in relation to step size is that, when signal averaging or other 'low-pass' filtering operations are performed, the step size of the output must be reduced by the square root of the effective number of points averaged; this should be obvious since the r.m.s. value of random noise decreases by the same factor.

Rule (2) is based on the standard Nyquist criterion. The reason for referring to noise rather than 'signal' is that, in any system, there will always be noise present up to the highest frequency of its electronic pass band whereas the true measured signal may be at a significantly lower frequency and, as for rule (1), noise cannot be properly averaged unless it is correctly digitized. Too low a digitization rate will result in an apparent increase in the noise level. In the case of Fourier transform (FT) instruments both noise and signal have frequency components close to the limit defined by the digitization interval; higher frequencies should be heavily attenuated by suitable electronic filters since, after Fourier transformation, such higher noise and signal frequencies will be aliased about the Nyquist frequency and appear as additional noise and spurious signals in the spectrum. A convenient way of ensuring compliance with rule (2) in many circumstances is to integrate the analogue signal between successive digitizations.

While rule (3) may be obvious it should be noted that, in most real situations, noise increases with signal. The relationship is square root in the well known case of shot (photon) noise. In other cases there is no strict relationship but there are a number of factors which, while they may be negligible in small signal situations (low signal : noise ratio), become dominant when the signal : noise ratio is very large. Examples are source instability, instrument vibration and air movement (schlieren) effects, all of which cause 'noise' directly proportional to signal. As a consequence of this the *precision* of digitization need not be as high as one part in n where n is the largest signal divided by the smallest increment as defined in rule (1). Wide dynamic range situations can therefore often be handled, without loss of information, by using some form of binary or decimal floating point system, typically performed by reducing the electronic gain prior to the analogue (A) to digital (D) converter in steps of two whenever an overflow situation is reached (with corresponding increases in gain as the signal decreases). In the author's experience it is very rare for the necessary precision of a single measurement to exceed 16 bits, *i.e.* 1 part in about 65 000.

1.1 What We Want

The basic purpose of computational operations is to convert the original spectroscopic data into a more easily understood and usable form whether it be a smoothed and flattened spectrum, a full qualitative/quantitative analysis, a determination of molecular/atomic constants or simply a 'yes'/'no' answer. The range of available techniques and algorithms is enormous and continuously increasing and many are covered in subsequent chapters. Here only some basic concepts, limitations and warnings are given.

Warning. Any computational operation will destroy some of the information contained in the original spectral data. Put another way, in generating the required information, other, ancillary information will be destroyed; the degree of relevance of this ancillary information will depend on the specific circumstances. A corollary of this is that computation can never generate information which was not implicit in the original data.

Only if the original data could be *exactly* regenerated after a computer operation would this statement not be true; even repeated conversions between transmittance and absorbance scales can noticeably degrade precision.

A simple example of the loss of ancillary information is the case of noise filtering; the 'noise' removed could be used to give some measure of the uncertainty of the spectral measurements and hence of the final results. In addition the 'noise' will be indicative of the relative spectral energy at different wavelengths/frequencies and may contain non-random features showing possible 'fault' conditions. This is illustrated by Figure 2 where (*a*) is the original infrared spectrum of polystyrene, (*b*) is the same after Savitzky–Golay[2] filtering and (*c*) is the difference (*a*) − (*b*). Examination of (*c*) shows at least four distinctive things.

1 The noise is higher in regions of atmospheric H_2O and CO_2 absorption indicating incomplete purging.
2 The general noise amplitude profile, highest at high and low wavenumbers, is typical of an FT instrument with a thermal detector.
3 Close inspection of the 1600 cm^{-1} region reveals features matching the first derivative of the water vapour spectrum in this region. This is typical of a calibration shift between the measuring of the sample and background spectra, possibly due to vignetting of the beam by the sample. In this example the effect was introduced deliberately by running the background with an open beam but with the sample slightly restricting the beam at the image of the Jacquinot stop. The Jacquinot stop is an optical aperture which serves to limit the divergence of the collimated beam in the interferometer: rays interfere at slightly differing wavelengths, depending on their angle to the optical axis, resulting in apparent shifts in the wavenumber scale.
4 The 'noise' increases in the regions of sample absorption. This is not true noise but represents the small distortion of the spectrum introduced by the filtering operations.

2 BASIC CONCEPTS

The enormous range of possible computational operations referred to earlier can be divided into three classes.

1 Manipulation; such things as filtering and reformatting.
2 'Simple' measurements, involving selected specific data such as peak positions and intensities.
3 'Complex' measurements, involving all, or a large part, of the full spectral data.

Manipulative operations include noise filtering, spectral subtraction, transmittance (T) to absorbance (A) conversion, generation of derivatives and other functions,

[2] A. Savitzky and M. J. E. Golay, *Anal. Chem.*, 1964, **36**, 1627.

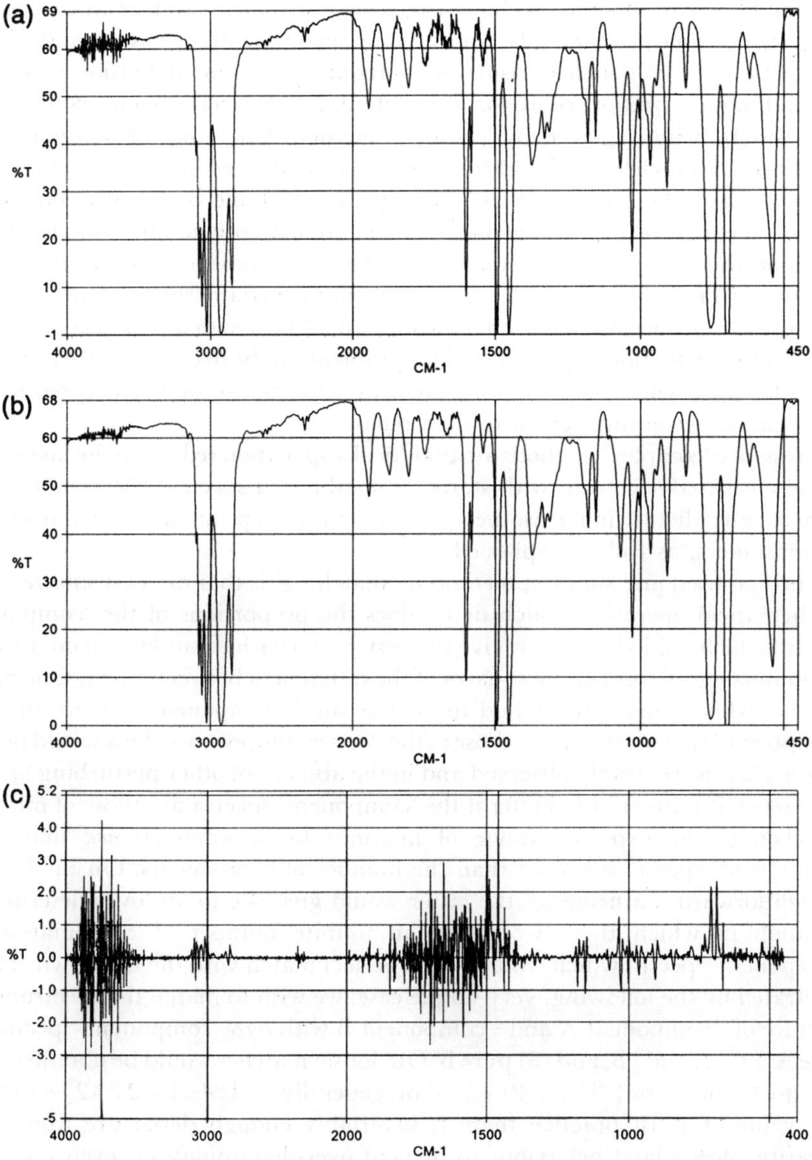

Figure 2 *Illustration of information removed by 'smoothing' a spectrum.* (a) *Original spectrum of polystyrene;* (b) *after Savitzky–Golay smoothing;* (c) *difference* (a) − (b).

background flattening and various scaling operations. They may be performed either as an aid to interpretation by the spectroscopist or as a 'pre-processing' operation prior to further computation. In the latter case heed should be taken of the warning above. For example, certain chemometric measurements on reflectance data have been found to perform better using the original data than the same data converted into the Kubelka–Munk format.

'Simple' measurements include the 'classical' single and multicomponent quantitative analysis methods, and certain spectral interpretation and library search routines. All such measurements require the prior definition of specific spectral points and/or rules for the definition of peaks. Such methods are increasingly being replaced by 'complex' methods where no, or minimal, prior selection of data is needed. In addition to obviating the necessity for selection, such methods have the advantage of using *all* the available data with consequent improvement in precision together with the ability to give an estimate of that precision from a single (spectrum) measurement. A possible disadvantage, if *no* selection of data regions is used, is that the inclusion of data from regions with little spectral significance and high noise, which would be avoided by a spectroscopist, may cause a reduction in precision. This problem can be overcome either by using some degree of selection or by using algorithms which recognize and reduce the significance (weighting) given to such regions.

In essence 'complex' methods for analytical application reduce to the 'matching' of the measured spectrum with stored or synthesized spectra. (Although it may seem a contradiction in terms, the actual matching step can be omitted in certain circumstances, as will be explained.)

The first used and simplest method of 'matching' is that of 'least squares',[3–5] a mathematical operation which determines the proportions of the 'component' spectra which, added together, give the best match with the unknown on the basis of minimizing the sum of the squares of the differences between them. One might assume, when using this method to analyse an N component mixture that the 'component' spectra should be those of the N pure components. This would be true if Beer's law were strictly observed and in the absence of other perturbing factors; in practice it is invariably better if the 'component' spectra are those of mixtures 'bracketing' the expected range of mixtures to be analysed *and* that more 'component' spectra are used than the number of components. On the basis of straightforward mathematics the latter would give rise to an 'over-determined' situation in which there would be an infinite number of combinations of 'component' spectra which would give a perfect match with the unknown. This is illustrated by the following, very simple case: we wish to match the spectrum of a mixture of $\frac{1}{2}$ component A and $\frac{1}{2}$ component B with *three* 'component' spectra (1) pure A, (2) $\frac{1}{2}$A and $\frac{1}{2}$B, and (3) pure B. Obviously matches would be obtained from $\frac{1}{2}(1)+\frac{1}{2}(3)$ or (2) or $\frac{1}{4}(1)+\frac{1}{2}(2)+\frac{1}{4}(3)$ or, generally, $x(1)+(1-2x)(2)+x(3)$ for any value of x. In practice there is invariably enough departure from strict linearity (Beer's law) behaviour to prevent over-determination (even noise will suffice) – the departures from linearity can be treated as if they represented additional components in the mixture. Figure 3 and Table 1 show an example of this effect in the case of mixtures of *ortho*- and *meta*-xylene.[6] Figure 3(a) shows the spectra of pure o- and m-xylene while (b) illustrates the clear difference between the 'synthetic' spectrum of a 50:50 mixture (generated by summing, in

[3] R. P. Bauman, *Appl. Spectrosc.*, 1959, **13**, 156.
[4] J. C. Sternberg, H. S. Stillo and R. H. Schwendeman, *Anal. Chem.*, 1960, **32**, 84.
[5] H. A. Barnett and A. Bartoli, *Anal. Chem.*, 1960, **32**, 1153.
[6] M. A. Ford and R. A. Spragg, 'Fourier Computational IR Spectroscopy', SPIE, 1985, vol. 553, 203.

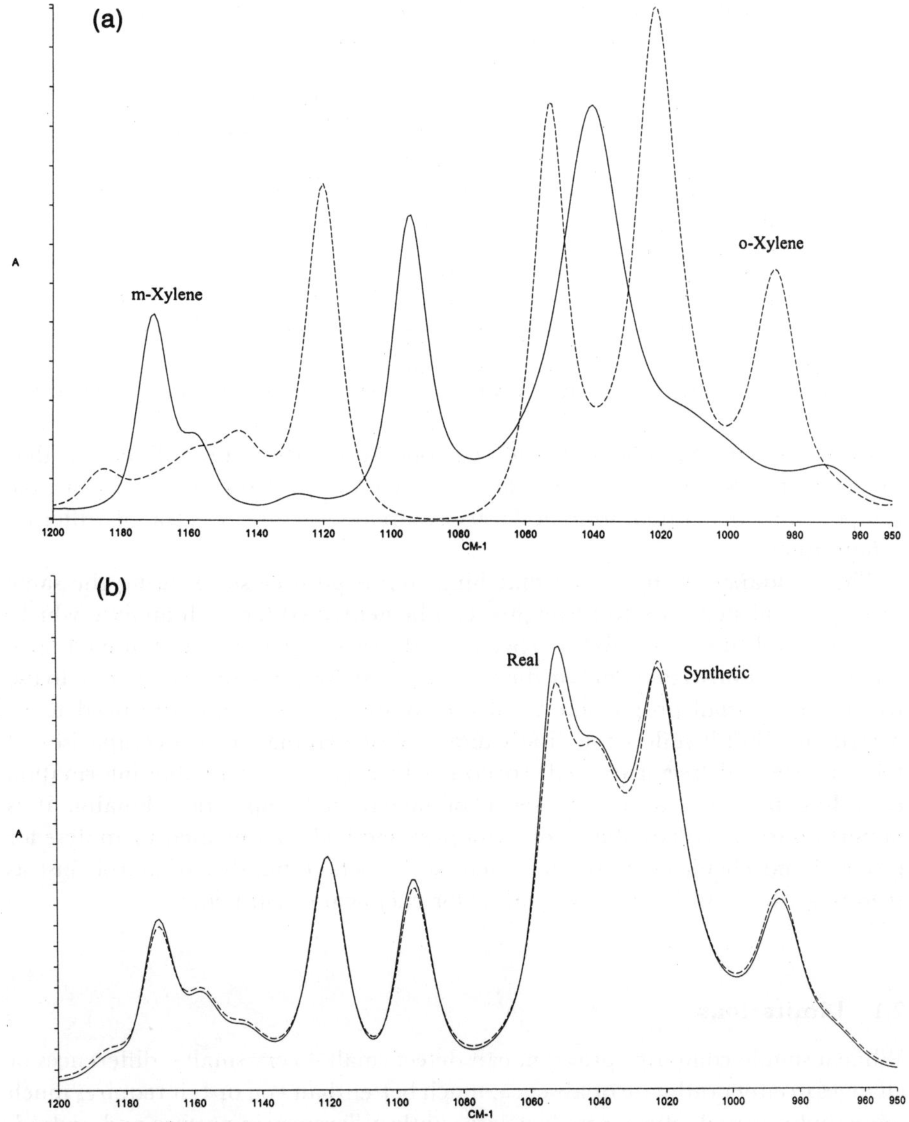

Figure 3 (a) *Spectra of* o- *and* m-*xylene*; (b) *Spectra of actual 50 : 50 mixture of* o- *and* m-*xylene (real) and computed from spectra in* (a) *(synthetic)*.

absorbance, the spectra of Figure 3(a) and halving the result) and the real spectrum of a 50:50 mixture. Table 1 summarizes the results of measurements with a series of mixtures and shows the steady improvement in precision as the reference spectra used for 'matching' go from the pure components through pure components plus one mixture to five standards (pure components + three mixtures). The causes of the non-linear behaviour can be chemical, physical or optical and the only requirement is that the non-linearity is not of the form of an

Table 1 *Analysis of 20, 40, 60 and 80% o-xylene in m-xylene*

	Average errors (%) 16 *measurements*	
Reference	*Absolute*	*Relative*
Pure components	4.2	13.6
Pure components + 1	2.3	7.5
Bracketing mixtures	1.6	5.6
Bracketing mixtures + 1	0.7	2.1
Five standards	0.12	0.3

equal scaling of the whole absorbance spectrum; this is virtually never the case.

In the more recent 'chemometric' type operations, the main difference is that the matching is not performed using real 'component' spectra but 'principle components', generated from a large group of mixture spectra of differing composition.

The avoidance of the actual matching step is possible since, using the same mathematical steps, a set of 'weights' can be generated for each analyte which, when multiplied by the mixture spectrum absorbance values and summed, give the concentration of that analyte directly. This can be quick and convenient to use when only a small proportion of all the components in a mixture need to be determined, but it suffers the disadvantage of preventing a direct comparison of the mixture and 'reconstructed' spectra, which can give valuable information regarding precision or the presence of unexpected impurities. Finally, it is important to appreciate that these 'complex' methods can be used to analyse for physical and chemical properties, such as the octane number of petrol, just as readily as they can be used to analyse for chemical constituents.

2.1 Limitations

While a simple computer program can detect small – very small – differences or changes in either ordinate or abscissa, much better than you or I, it requires much more sophisticated software to deal with subtle differences in pattern and, indeed, to date there are still many areas in which the computer is no match for the skilled spectroscopist. This situation can be expected to change over the next few years as these spectroscopic skills become incorporated in 'expert systems', computer jargon for programs which make decisions/produce results based on itemized pieces of 'knowledge' stored as 'rules' within the program. In the long term such systems should have the potential to outperform even the best spectroscopist simply because they can be based on the knowledge of not one but many, and bring to bear *all* that knowledge simultaneously. The limitation is mainly one of resource and cost rather than concept and it may be some years before it becomes economic to tackle many of the potential opportunities.

2.2 Where We Want

When a spectrum is generated, in all probability the data are needed, not at the spectrometer, but at a convenient desk terminal. Equally, access to stored spectra is needed at a terminal, not at your or someone else's computer archive. Within one laboratory or organisation both results can easily be achieved by using any standard data format and any local area network. However, between different or remote organizations, it is necessary to have mutually usable links and data formats. Links are no longer a major problem as systems, such as 'gateways' or 'bridges', are readily available for cross connecting the popular local area networks (LAN) (*e.g.* Ethernet and Token ring) and modern spectrometers usually have in-built or optional LAN ports. Data formats can still be a problem as, until recently, the individual instrument manufacturers each used their own and only now are starting to provide optional conversion into one of the accepted standard formats (JSCAMP or Net CDF). Even with these a potential problem still exists, namely that of abscissa data intervals for which JCAMP provides free choice; to compare spectra properly they must use (nominally) identical abscissa points. The problem can potentially be overcome by interpolation to a new interval but such interpolation may involve significant loss of precision. Excellent interpolation algorithms do exist which come close to the ideal, namely the ability to interpolate to a new, but similar, interval and back, reproducing exactly the original data. The whole question of standard data formats is covered more fully in Chapter 4.

2.3 When We Want

The issue of data storage and retrieval is covered extensively in Chapter 4 and 5. The only general question which will be mentioned here is that of *what* to store rather than how it is stored. The rapidly increasing availability and reducing cost of bulk storage is being matched by the increasing output of modern spectrometers which can generate prodigious quantities of data in a short time. It will probably remain impractical to store all this 'raw' data,[7] although it has to be done where demanded to meet legal requirements and judgement will continue to be needed as to whether it is sufficient to keep only the final result or some part processed information. The quandary is that storing all the 'raw' data gives the option of subsequent reprocessing using other algorithms while any data reduction carries the risk, as stated earlier, of losing relevant information.

2.4 How We Want

The question of the method of presentation of the results of computer data processing is far from trivial. Like the processing itself it can be divided into

[7] This view has been challenged by others who believe that all data, optimally compressed, can be stored economically. The author still questions this in the case of certain time resolved data such as high (chromatographic) resolution GC-IR runs.

methods associated with manipulative operations and those for presenting the results of simple or complex measurements.

As stated before, manipulative operations are primarily to aid the spectroscopist in the interpretation of the data and the methods of presentation form an important, integral part of this. The use of colour and 'windows' type environments has greatly enhanced the ease with which whole spectra, or selected regions of single spectra can be examined and compared while, for multiple time or spatially resolved spectra 'stacked plots' or coloured 'intensity maps' are extremely valuable. The stacked plots can be further enhanced by the ability to rotate so as to 'view' from any direction, while the ability to view any section through an intensity map is very valuable. No doubt further improvements in these areas are still to come.

The presentation of 'answers' – the results of simple or complex measurements – has been developed extensively for Laboratory Information Management Systems (LIMS) and, although these were originally based mainly on chromatographic data, the concepts are equally applicable to spectroscopic data. Results can be presented, for example, as a full analysis report, incorporating measurements from a range of analytical operations, in any required format, to meet specific user needs. Typically, such reports will include other relevant data such as date/time, operator, sample origin *etc.* and the complete system will include provision for archiving all data, as specified to meet any 'good laboratory practice' (GLP) or similar quality system or accreditation requirements, as well as having security systems to limit unauthorized access to the results.

On a broader front, it is fast becoming the rule that spectroscopic data can be transferred into any of the standard commercial spread sheets and word processing/desk top publication packages such as Lotus 1-2-3 or EXCEL. Not only does this provide the user with an almost infinite range of presentation options but it also permits direct incorporation of results into text for publication.

2.5 Reliability

The general issue of valid analytical measurements is of very great and increasingly recognized importance. All computational operations involved in obtaining the final analytical result inherently form a part of the system which has to be verified. This is far from straightforward as simply verifying the basic algorithms used is not sufficient; it must be confirmed that the total computational program performs correctly at all times under all conditions – that is with all possible values of input parameters. (The author has experienced programs that will operate correctly for days and then start producing spurious results or that operate correctly for all except a very limited set of input parameters.) Rigorous testing of all possibilities is clearly impossible and work is being done in many organizations at present to try and establish practical, workable systems and regulations for software checking.

The problem can be helped significantly by having these requirements taken into account in the writing of the software but relatively little has been done to date in this respect. In addition the complete computer system can greatly aid in

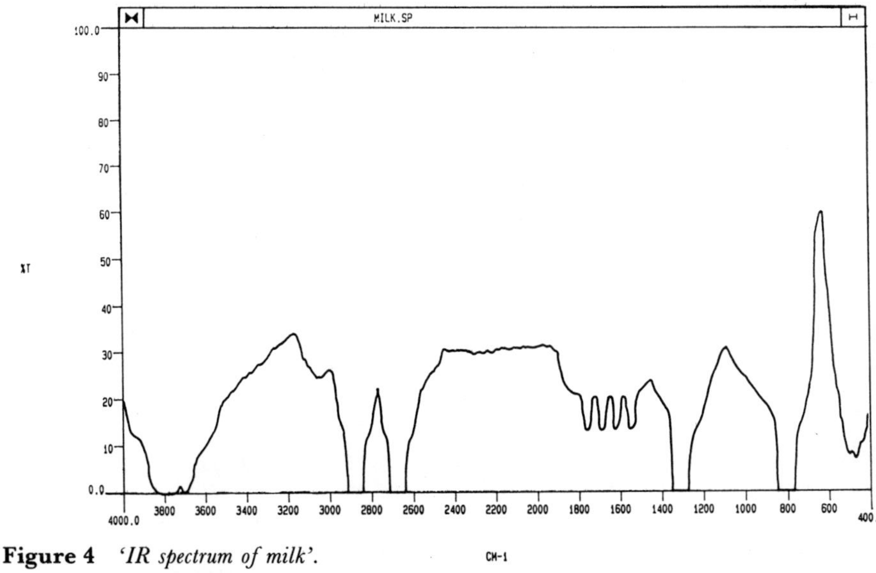

Figure 4 *'IR spectrum of milk'.*

meeting the requirements imposed by standards such as ISO9000, GLP *etc.*, by keeping records of such things as time since last service/ standardization *etc.* Some systems already do this and it can be anticipated to become the rule rather than the exception in the future.

3 CONCLUSION

This chapter has attempted to outline the enormous range of existing and potential applications of computers to spectroscopic data. However, computers still cannot do everything and this is lightheartedly illustrated by the 'infrared spectrum of milk' Figure 4. No known or projected computer could interpret this; no one knows how to give a computer a sense of humour!

CHAPTER 2

Data Manipulation in Spectroscopy

A. S. GILBERT

1 INTRODUCTION

Data manipulation deals with the realm of operations lying between successful acquisition of data from the spectrometer and the extraction of information, that is to say those processes that convert raw data into forms more suitable for quantitative analysis, measurement of peak positions and other operations that gather facts as a prelude to gaining physicochemical knowledge about the system under study.

Data acquired from a spectrometer consist of a spectrum of intensity values belonging to channels separated either by equal frequency proportional intervals (wavenumbers) or equal wavelength proportional intervals (angstroms, nanometres or microns). In the simpler category of data manipulation to be discussed the data channels remain independent; thus an output datum in any particular channel is a function of input data in that channel only. Transmittance to absorbance is manipulation of data that belongs to this category. Only one set of data is output, but more than one set of data may be input, for example when subtraction is performed.

The more complex category is represented by the operation of smoothing data to enhance signal-to-noise ratio (SNR) or sharpening spectral bands to enhance resolution. Here the data channels do not remain independent; any output datum is a function of input data from many channels but with a heavy bias towards channels that are adjacent to each other and to the output. This achieved *e.g.* by convolution of the input data set with a triangular function (in the case of smoothing) to generate one output set of data.

The process may be linear; *i.e.* involving a single and ideally reversible operation, or alternatively it may be iterative and non-reversible, an example of the latter kind being deconvolution of band shape utilizing the 'principle of maximum entropy' which will be discussed later.

2 DATA PROCESSING INVOLVING NO MOVEMENT OF DATA BETWEEN CHANNELS

2.1 Transmission to Absorption

IR and UV–VIS spectrometers measure the transmittance, T, which is the fraction of radiation remaining after passage through the sample. This quantity is not proportional to sample concentration or path length but can be converted into one that is, by the relation

$$A = \log_{10} 1/T \tag{1}$$

where A is the absorbance.

Most if not all modern UV–VIS spectrometers do this automatically but it is normal for the operators of Fourier transform infrared (FTIR) spectrometers to initiate this conversion themselves by using computer software.

The statement of the linear relationship of absorbance to sample parameters such as concentration is called Beer's Law (although it should perhaps be renamed after Bouguer and Bernard[1]). However, deviations from this law always occur to a greater or lesser extent. These are usually owing to physicochemical effects, which will not concern us here further, but also can arise from settings of instrumental parameters. For instance where the actual width of a band is of the order of or less than the spectrometer's slit function, both the band height and area will be significantly modified by differing and non-proportional extents depending on the absorption strength. This is because the transmission profile (or shape) of the spectrum varies with sample concentration and path length due to its logarithmic relationship with the absorbance and will therefore be convoluted (see below) in a different way with the slit function. Varying the slit function from experiment to experiment introduces even more deviations. [*N.B.* the effective slit function in an FTIR spectrometer may be a combination both of the 'physical' resolution (maximum path difference) and a software apodization function (implemented transparently to the user).]

Apart from these problems with Beer's Law, conversion into absorbance of spectral regions of low transmittance will lead to roughly exponential increases in noise compared to other parts. This is principally owing to the noise being either virtually independent of transmittance (IR) or proportional to $\sqrt{T}$ (UV–VIS), but also can come from lack of dynamic range during digitization.

These and other considerations normally limit accurate quantitation using IR absorbance bands to about 1.5 a.u., as long as spectral subtraction is not employed. In the latter situation the limit can be somewhat lower, but it is possible to rely on UV–VIS bands of up to 3 a.u. or so providing they are at least an order of magnitude broader than the instrumental slit width and the instrument has a low stray light specification.

Measured Raman spectral intensity is, like absorbance, proportional to sample concentration, other things being equal.

[1] D. T. Burns, in 'Advance in Standards and Methodology in Spectrophotometry', ed. C. Burgess and K. D. Mielenz, 1987, pp. 1–19.

2.2 Spectral Subtraction

Raman and most FTIR spectrometers are single beam devices and as vibrational spectra of solvents are commonly rich in bands, spectral subtraction routines are a central part of any associated data system. Most manufacturers of commercial systems implement these routines in a variety of ways, the user having the choice of doing a single subtraction with fixed constants, or doing the process interactively whereby the subtraction constant is varied up or down in small amounts until a satisfactory end result is achieved. Many data systems also possess an automated subtraction mode where if the user is confident that a particular spectral region contains only solvent bands s/he can instruct the computer to adjust the subtraction constant until the region is flattened, with the hope that complete 'cancellation' of all solvent bands across the spectrum will occur.

Spectral subtraction is performed on absorbance data for IR and UV–VIS spectroscopy. Removal of interfering solvent bands is probably the most common application but removal of 'impurity' bands using a 'pure impurity' reference spectrum as subtractant is probably just as popular. Well prepared alkali halide discs often yield sufficiently good concentration-proportional IR spectra so that spectral subtraction can be carried out with a fair degree of confidence. This enables quality control to be performed on solid samples by subtraction of the reference spectrum of a known *pure* sample from, say, a production batch of the same compound to look for possible impurities. Such a process will not always work for solids as the method of preparation of discs and the alkali halide matrix itself can induce sample-to-sample spectral variation even if polymorphism does not occur. Other techniques such as diffuse reflectance and photoacoustic absorption may overcome these difficulties.

But as with all types of data manipulation, care must be exercised in case the conditions are inappropriate for the process. Subtraction always increases the noise across the spectrum and increases can be very dramatic in regions where highly absorbing bands are present. These are precisely the circumstances where Beer's Law deviations become significant and such conditions can easily lead to the generation of artefacts with the danger of them being mistaken as bands. Restrictions must therefore apply and as usual these are greatest for IR absorption spectra where spectral regions of more than about 1 a.u. should be avoided if the purpose is accurate quantitative analysis. Greater absorbances can be tolerated for UV–VIS absorption and strong Raman bands generally cause fewer problems because their associated signal dependent photon statistical noise fortunately only increases as the square root of the signal strength. But Raman spectra can, like IR, suffer from band shape changes brought about by physicochemical interactions and it should be noted that Raman detectors (both P.M. tubes and arrays) can suffer from saturation so that very strong bands are depressed in apparent intensity.

2.3 Baseline Correction

Ideally the spectral trace should not possess any deviations from the horizontal other than those from the bands of interest, *i.e.* IR or electronic absorbtions in the case of IR and UV–VIS spectra and vibrational scattering in the case of Raman spectra. However, the latter are often underlain by broad fluorescence emission lowering SNR and solid phase IR spectra are commonly affected by scattering which reduces transmission towards higher frequencies, again lowering SNR and also restricting dynamic range. These problems can only be corrected properly by attending to the sample but data manipulation is appropriate where the effects are not too severe and where display of the spectrum would result in the peaks of interest being small and 'squashed'. It is also necessary if peak positions are to be accurately measured where they lie on sloping baselines or where the spectrum is to be compared point-by-point with others, for instance in a library matching procedure, to find an unknown. Care should be exercised however so that broad bands are not assigned to fluctuations in the background. This is particularly important for IR spectra from severely scattering alkali halide discs where broad absorptions, for example from N–H band stretching, can be lost in the rapidly falling transmission curve.

Baseline correction is performed by ratioing transmission data or subtracting from absorption or Raman spectra an artificially constructed background. While it might be expected that the background of transmission spectra from solid state samples should follow some simple or known scattering function, the distribution of particle size and shapes causing the effect is too complex, although parabolic curves are sometimes a reasonable approximation. Most commercial software usually offers both parabolic and linear functions which can be applied in an automatic mode but bearing in mind the caveat in the previous paragraph, it is better for operators to use their own judgement and construct a baseline of their own by selecting suitable points throughout the spectrum and 'joining' them up so that polynomial interpolation can be implemented to give a smooth curve.

3 LINEAR DATA PROCESSING INCLUDING DATA TRANSFERS BETWEEN CHANNELS

3.1 Smoothing and Differentiation

Bands in spectra have breadth and are commonly overlapped with each other. Many factors including molecular collisions, uncertainty processes and finite 'slit' widths conspire to blur ideally sharp peaks and in addition spectra are degraded by noise resulting from both the statistical (photonic) nature of the signal and the thermal background of the detector and measurement system. Therefore in order to extract the desired information from a spectrum it may be necessary to reduce or remove (deconvolute) breadth and/or noise.

Differentiation and smoothing are well known data manipulation methods; the former is extensively used to resolve or separate bands in UV–VIS spectroscopy, the latter is commonly applied to IR and Raman spectra on account of the greater noise manifested in these techniques. Savitzky and Golay[2] have developed convolution functions for both these operations; these functions are essentially

[2] A. Savitzky and M. Golay, *Anal. Chem.*, *1964*, **36** (8), 1627.

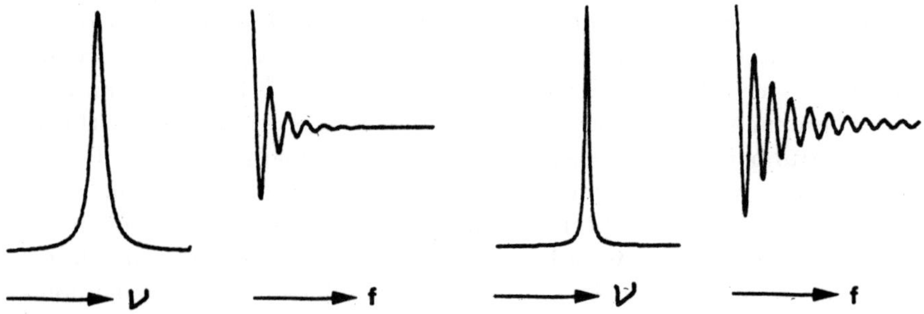

Figure 1 *Lorentzian bands and their respective Fourier transforms*

filters which are passed over the spectrum, the convolution operation being described below.

$$A'_v = \sum_{x=-m}^{x=m} F_x A_{v+x} / F \qquad (2)$$

In fact any roughly triangular (convex) function can be used to smooth data but Savitzky–Golay functions duplicate the process of fitting adjacent points to polynomials by least-squares regression. A set of coefficients for a typical Savitzky–Golay smooth are $F_x = -3$, 12, 17, 12, -3 where $m=2$ and $F=35$. A simple triangular smoothing function could be $F_x = 1$, 2, 1 where $m=1$ and $F=4$.

The outcome of smoothing is to improve SNR, thereby allowing better estimations of peak positions, intensities, *etc.*, but this is always achieved at the expense of reducing band height and increasing band width (thus reducing resolution). The amount by which these parameters change depends on the relation of the width of the smoothing function to the spectral bandwidths;[3] thus the relative band heights of differently shaped bands in a spectrum change, but fortunately the areas under bands are unaffected so that there is still a means of quantitation available. This is important if the noise characteristics of a set of spectra being compared require different degrees of smoothing. The optimum increase in SNR is achieved by smoothing with a function that exactly matches the line-shape (matched filter) but this is not, of course, possible if the band shapes differ.

Differentiation always decreases SNR and has therefore chiefly been employed on UV–VIS absorption spectra with their typically high SNRs. Even-order differentiation does produce band-like profiles but relative band area relationships are lost and side-lobes are generated. Broad bands with their gentler slopes tend to be suppressed in relation to sharp bands but analytes often yield narrow bands whereas other components in mixtures or solvents often given broad sloped backgrounds so that differentiation offers some advantages in these situations.

Both smoothing and differentiation are part of the general process of convolution. This is best considered by examining the data in the Fourier domain. Figure 1 shows the FTs (only the cosine coefficients are shown in this and

[3] C. G. Enke and T. A. Nieman, *Anal. Chem.*, 1976, **48** (8), 705A.

succeeding figures – that is they are assumed to be symmetric about the origin) of Lorentzian bands. Initially the waves decay exponentially from the origin: the broader the spectral band the more rapid the decay. It is helpful to think of the Fourier domain as a frequency domain (although this designation is arbitrary); sharp spectral bands therefore yield oscillations with many more high frequency components (waves) in the Fourier domain (data appearing much further from the origin) whereas broad spectral bands give waves concentrated much nearer the origin (rapid decay). According to the convolution theorem, convolution in one domain is equivalent to a point-by-point multiplication in the other domain, the multiplying or transfer function being the FT of the convoluting or filter function. (See p. 31 for further discussion of convolution.)

Figure 2(*a*) shows the Fourier domain of the Lorentzian doublet displayed as (*b*). The effect of addition of noise to this doublet is shown in (*c*) and (*d*). The transfer function for matched filtering is Figure 2(*e*) and point-by-point multiplication of this with (*c*) will tend to squash the signal towards the origin of the domain while severely reducing the noise elsewhere. The result, illustrated in Figure 2(*f*), is that SNR in this particular case has improved by almost 5.5 times [compared to (*d*)] but at the cost of a certain loss of resolution.

An alternative transfer function for smoothing is a boxcar truncation as shown in Figure 2(*g*). As can be seen this function retains unaltered all the signal components until they disappear under the noise after which it simply eliminates everything to higher frequencies. As shown in Figure 2(*h*) this allows spectral resolution to be largely retained but noise reduction is not so good, the improvement in SNR over Figure 2(*d*) being just under three.

Finally Figure 2(*i*) shows the transfer function for a 'cubic-quartic' Savitzky–Golay filter having about twice the number of points as the full-width at half height of the individual Lorentzians. This condition gives the optimum SNR gain for these filters and the result in Figure 2(*j*) is more than three times better than (*d*).

Savitzky–Golay filters may be considered as general purpose functions for use when little is known about band shapes. They have the additional merits of being very easy to implement and are widely available in commercial software.

Thus significant gains may be obtained by smoothing spectra, which could otherwise only be achieved by greater sample concentration, signal averaging or running the spectrometer more slowly to improve SNR experimentally. Particular circumstances will normally dictate which course to follow but it should be remembered that the time factor is roughly proportional to the square of the desired gain in SNR if resort is made to additional measurements: *e.g.* nine successive spectra are needed for co-addition to improve SNR by three times.

3.2 Resolution Enhancement

Figures 3(*a*) and (*b*) show a relatively noise-free Lorentzian doublet and its corresponding FT, respectively. The transfer function shown in (*c*) is for second differentiation (inverted) and the result is displayed in (*d*). The components of the doublet have been separated but overall band shape is not very satisfactory with negative side-lobes. This transfer function rises from zero at the origin of the

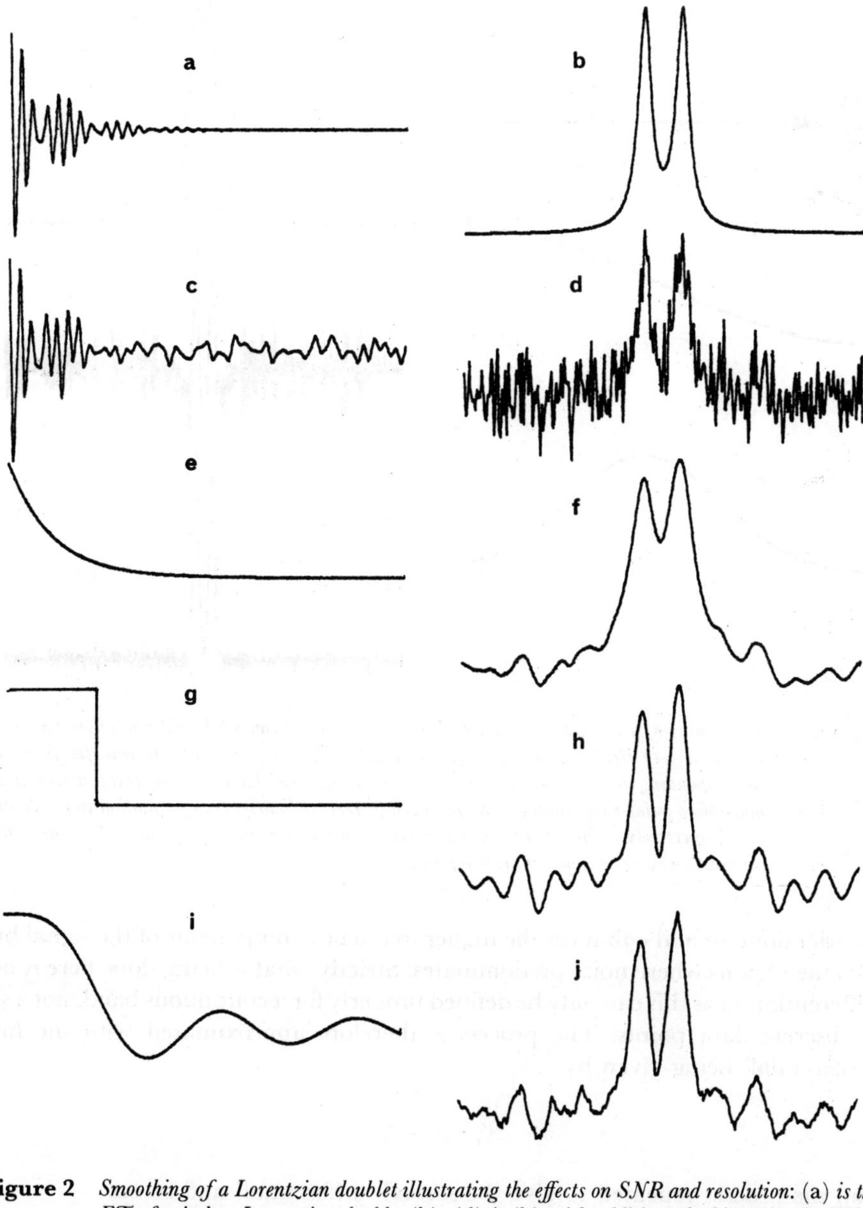

Figure 2 *Smoothing of a Lorentzian doublet illustrating the effects on SNR and resolution*: (a) *is the FT of noiseless Lorentzian doublet* (b); (d) *is* (b) *with addition of white noise the FT of which is* (c) *showing how the noise is spread out across the Fourier domain*; (e), (g) *and* (i) *are transfer functions which when multiplied point-by-point with* (c) *followed by reverse Fourier transformation yield* (f), (h) *and* (j), *respectively. Note that in practice,* (f) *and* (j) *would be obtained more conveniently by convolution with the corresponding filter functions (a matched Lorentzian or Savitzky–Golay function, respectively) without recourse to Fourier transformation*

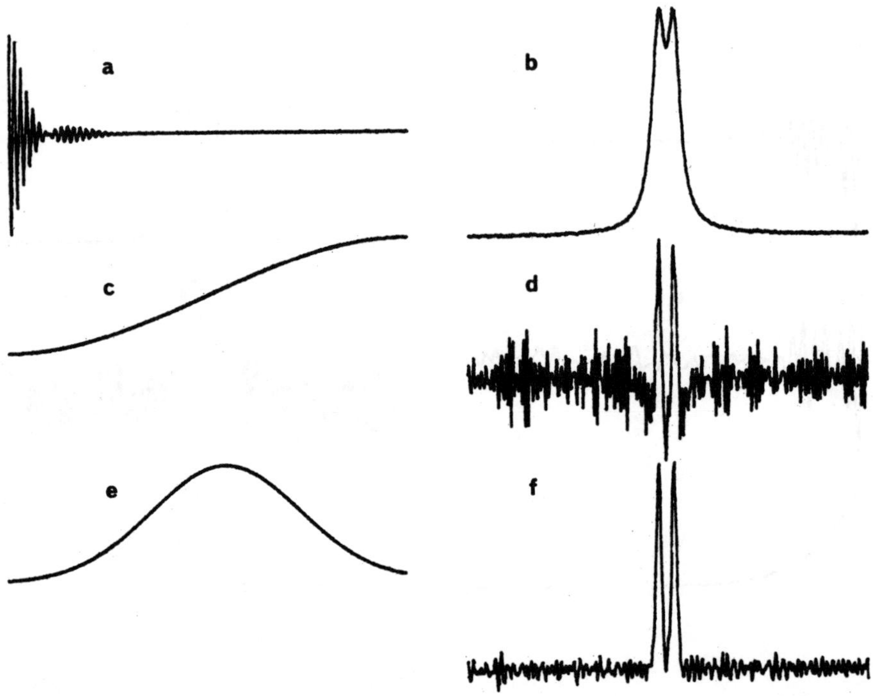

Figure 3 *Sharpening of a Lorentzian doublet illustrating the effects on SNR and resolution*: (a) *is the FT of a very slightly noisy Lorentzian doublet* (b); (c) *and* (e) *are transfer functions corrresponding to second differentiation (inverted) and LDGA (see text) which when multiplied point-by-point with* (a) *followed by reverse Fourier transformation yield* (d) *and* (f), *respectively. Note that in practice* (d) *is more conveniently obtained by convolution without recourse to Fourier transformation*

Fourier domain and enhances the higher frequency components of the signal but also the regions where noise predominates. Strictly what is being done here is not differentiation as this can only be defined properly for a continuous band, not a set of discrete data points. The process is therefore approximated with the first 'differential' being given by

$$A'_{v} = A_{v} - A_{v-1} \tag{3}$$

and the process repeated to give the second 'differential' and so on.

A more elegant method of sharpening the bands it to match the deconvolution to the band shape. This is the basis of the so-called Fourier resolution enhancement (FRE) or Fourier self-deconvolution (FSD) methods. For instance, the Fourier domain signal of a Lorentzian band decays by $\exp(-f)$ so that multiplication by the transfer function $\exp(f)$ gives a set of pure cosine waves which when transformed back to the spectral domain yield an infinitely sharp line (Figure 4).

Sadly such an ultimate degree of sharpening cannot be achieved in practice because of the noise which is ever present in the data set, if not from the detector

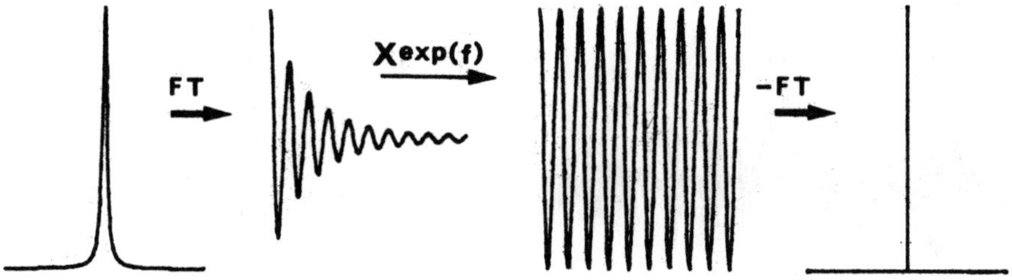

Figure 4 *Complete deconvolution of Lorentzian lineshape (unrealizable in practice)*

then by virtue of the finite word length of the computer used to contain it. This noise will usually be distributed roughly equally across the Fourier domain, (*i.e. white*). The effect of multiplication by exp(f) vastly increases the higher frequency components and because noise in any one channel in the Fourier domain appears in *all* channels in the spectral domain, the result is that the spectral bands are swamped.

Fortunately the problem can partially be alleviated by apodization, that is using a second function to reverse the exponential rise of the first in the high frequency region. There is a penalty however in that the apodization function imposes its own shape on the final band. This is illustrated in the case of Gaussian apodization in Figures 3(e) and (f). The net result is that a useful degree of resolution enhancement is obtained with control of noise, and this is brought about because the overall 'weighting' of the combined transfer functions enhances the signal in the middle frequency regions of the Fourier domain whereas the high frequency region, where noise predominates, is suppressed. This transfer function can therefore be thought of as a frequency band pass filter, whereas smoothing and differentiation have transfer functions that are respectively low and high pass frequency filters.

Both the degree of deconvolution and apodization can be separately controlled using FRE (FSD). Weakening the apodization gives narrower bands but at the expense of higher noise [Figure 5(a)] while over deconvolution generates negative side lobes [Figure 5(b)].

The combination of Lorentzian deconvolution Gaussian apodization (LDGA) has been extensively utilized[4] for the resolution enhancement of proton NMR spectra for many years and is particularly suited to this technique because of the generally consistent and fairly pure Lorentzian line-shape of the observed bands. Multiplets can be revealed and coupling constants measured that would not otherwise be detected without recourse to much higher magnetic fields.

The trade-offs between reduction in bandwidth R and reduction in SNR are given in Table 1 for the LDGA transfer function which demonstrates that apodization only controls the noise within limits.

Alternative apodization functions can be used[5] which do not allow the noise to expand so quickly for a given reduction factor but this is at the expense of broader

[4] J. C. Lindon and A. G. Ferrige, *J. Magn. Reson.*, 1981, **44**, 566.
[5] J. Kauppinen, D.J. Moffat, H. H. Mantsch and D. G. Cameron, *Appl. Spectrosc.*, 1981, **35**, 271.

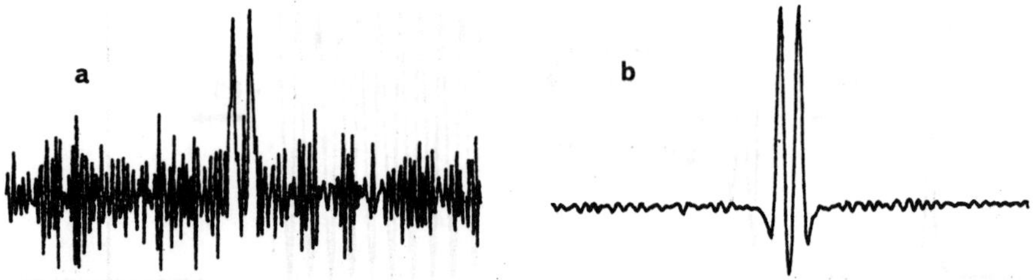

Figure 5 *Effects of varying the degre of apodisation and deconvolution when using the LDGA function in conjunction with the Lorentzian doublet* 3(b): (a) *weak apodization whereby the transfer function is allowed to peak to a higher frequency than in* 3(e); (b) *over-deconvolution by using a rising exponential appropriate to a broader Lorentzian band*

Table 1 *Trade-off factors*[a]

R	*SNR*
1	1.6
0.8	0.94
0.6	0.35
0.4	0.026
0.3	7.9×10^{-4}
0.2	1.4×10^{-7}

[a] Figures are for a completely deconvoluted Lorentzian being converted into a pure Gaussian. They will vary slightly depending on the exact width of the band.

wings to the final bands and anyway, whatever type of apodization used, noise increases in an exponential manner for *R* factors below about 0.25.

Gaussian bands when deconvoluted require sharper apodization to counteract the power two exponential deconvolution function. The degree of narrowing possible is smaller than for Lorentzian bands because the noise is harder to control.

As long as the transfer function varies smoothly from a value of one at the origin, then the area under bands remains constant, whether they are subjected to broadening or narrowing. If side-lobes are generated (from over deconvolution) the area under negative lobes subtracts from the positive regions to equate with the original band area. This area constancy is a consequence of the fact that the value of the interferogram at the origin is equal to the integrated area under the spectrum.

A consequence of the crudely exponential dependence of noise on width reduction by deconvolution as shown above, is that it is usually better to change the operating resolution of the spectrometer, if possible, when significant band sharpening is required and when the peaks are not 'blurred' by intrinsic processes. This is because the experimental trade-off in SNR is roughly the square or a bit more of the increase in resolution and thus *not* exponential.

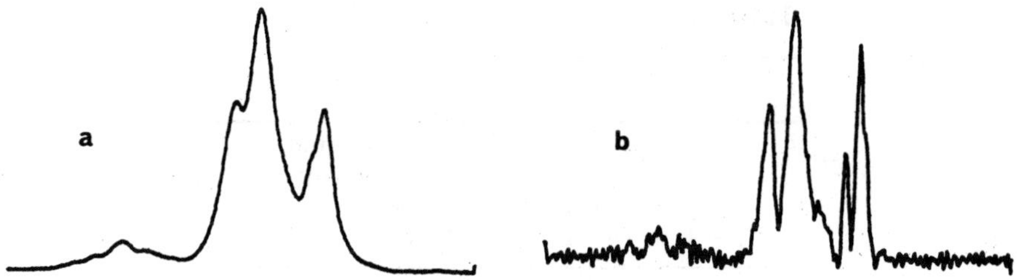

Figure 6 *LDGA resolution enhancement of the C–H stretching region of a hydrocarbon chain containing molecule*: (a) *original spectrum*; (b) *final spectrum*

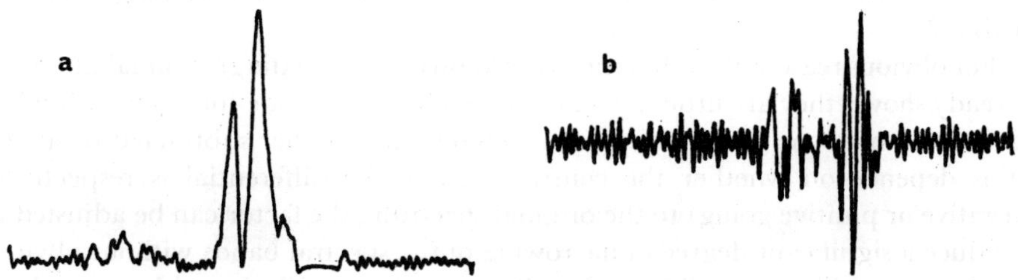

Figure 7 *Other manipulations on an IR spectrum*: (a) *removal of the rightmost doublet of* 6(b) *followed by further LDGA resolution enhancement*; (b) *second differential (inverted) of* 6(a)

The treatment by LDGA of the IR C–H stretching region [Figure 6(*a*)] of a molecule containing a long chain hydrocarbon is shown in Figure 6(*b*) demonstrating separation of the methyl and methylene bands. Measurement of the relative areas under the rightmost doublet allows the ratio of methyl to methylene groups to be estimated by comparison with resolution enhanced spectra of 'standard' hydrocarbons.

The lineshapes of the individual bands in this example are approximately Lorentzian and variable in width. However variable width does not necessarily prevent more bands being separated, for the doublet described above can be stripped off and further deconvolution carried out as shown in Figure 7(*a*). If this were not done the further treatment would cause over-deconvolution of these bands.

Finally Figure 7(*b*) illustrates the second differential (inverted) of Figure 6(*a*); it is evident that this is far inferior to the result of LDGA. Note however that the heights of the second differential peaks depend on the sharpness of the original bands, and they are unrelated to the strengths of the bands. A sharp small spike from an electrical disturbance for example would produce a very substantial derivative spectrum!

Many spectra have band shapes however that are not as amenable as the examples described above, and this is particularly true for UV–VIS spectroscopy. For these cases some type of general purpose deconvolution functionality is needed rather than the specific LDGA process and this can be achieved by adapting differentiation as described by equation. (3).

Table 2 *Differentiation functions*

F_x	F	*Process*
−1, 1	1	1st differential
1, −2, 1	1	2nd differential
−1, 3, −3, 1	1	3rd differential
1, −4, 6, −4, 1	1	4th differential

This equation can be implemented by a convolution function where $F_x = -1, 1$ $(x = 0, 1)$. Repeated applications to produce the higher order differentials can be achieved instead by single applications of the convolution functions shown in Table 2.

For obvious reasons these functions are known as alternating binomial filters. As already shown they are little use on their own because of side-lobe generation but if, for instance, one of the even-order differentials is either subtracted or added (this depends on whether the central peak of the differential is respectively negative or positive going) to the original spectrum, the factor can be adjusted to produce a significant degree of narrowing of the spectral bands without obvious production of side-lobes. This can be followed up by convolution with a smoothing function (say a simple non-alternating binomial) to suppress noise and the overall process is equivalent to using a frequency band pass transfer function as in Figure 3(*e*). The second-order differential is probably most suitable for wide tailed bands like Lorentzians but higher (even) order differentials are required to sharpen bands of narrower (more Gaussian like) profiles. A recent publication by Jones[6] demonstrates the results of applying these binomial filters to UV spectra of pharmaceutical formulations for the purpose of improving the precision and accuracy of quantitative assays.

4 MAXIMUM ENTROPY METHODOLOGY (MEM) FOR RESOLUTION ENHANCEMENT OF SPECTRA

Deconvolution (including differentiation) suffers from two basic defects. Firstly the noise is not separated from the deconvolution process whereas spectra only acquire noise during measurement, *i.e.* after the bands have been convoluted with a blurring function and have acquired their breadth. The detrimental effects on SNR are a consequence of this. Secondly, as demonstrated in Figures 3(*d*), 5(*b*) and 7(*b*), negative values can be produced and this is plainly not sensible for instance in the case of an emission spectroscopy such as the Raman effect.

These deficiencies can be addressed by deconvoluting in an iterative fashion with utilization of the 'principle of maximum entropy'. If a guess is made at the true spectrum, the object, and this trial object is blurred and the result subtracted

[6] R. Jones, *Analyst*, 1987, **112**, 1495.

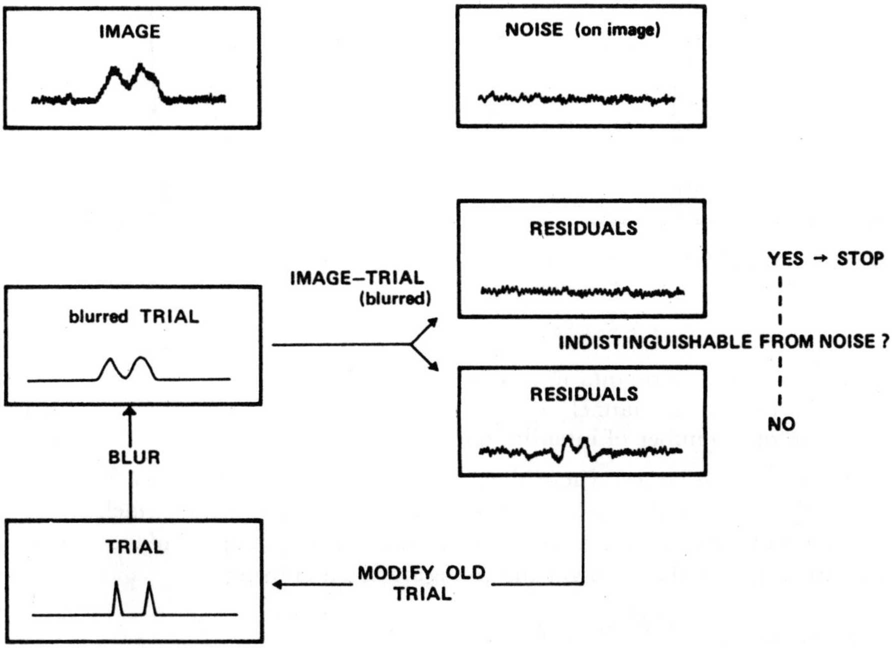

Figure 8 *Iterative deconvolution*

from the measured spectrum, or image, the residuals will, unless one is very lucky, be plainly distinguishable from the noise itself assuming of course that the overall noise characteristics are known. The residuals, however, can be utilized to modify the trial object with the proviso that no values are allowed to be negative and this new trial can then be blurred and new residuals computed and so on until a set of residuals is obtained which cannot be distinguished from the noise. This iterative procedure is illustrated in Figure 8.

The trial object at this point is however not necessarily the object itself. This is because there are very many sets of residuals that are indistinguishable from the noise (which by definition is only known by its overall characteristics, *e.g.* root mean square value, frequency distribution, *etc.*) and therefore very many compatible trial objects, that selected by the iterative cycle being an arbitrary outcome of the initial guess and the mechanics of the procedure.

Fortunately it is possible to make a logical choice as to which candidate is most likely to be the real object. If a spectrum is regarded as a finite number of intensity quanta (*e.g.* photons) divided among a number of channels then a figure for the likelihood of that spectrum can be given by the ways W that the quanta can be arranged among the channels.

$$W = \frac{\mathcal{N}!}{(\mathcal{N}-n_1)!n_1!} \cdot \frac{(\mathcal{N}-n_1)!}{(\mathcal{N}-n_1-n_2)!n_2!} \cdot etc. = \mathcal{N}! / \prod_{\nu=1}^{\nu=M} n_\nu!$$

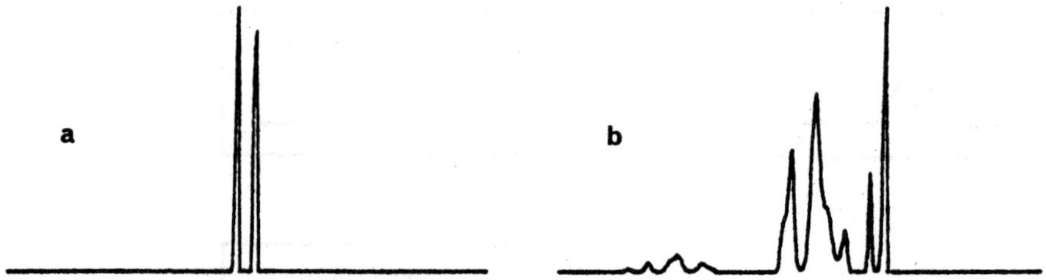

Figure 9 *MEM deconvolution for resolution enhancement*: (a) *utilising the Lorentzian doublet* 3(b); (b) *utilising the IR spectrum* 6(a)

where n_v = number of quanta in channel v
M = number of channels
N = total number of intensity quanta

The likelihood is largest where all n_v are equal (*i.e.* the spectrum is a featureless straight line) and smallest where all the quanta are in just one channel. A more useful form of the equation is obtained by using Stirling's approximation (where there are large numbers) to get the configurational entropy

$$S = \log_n W = -K \cdot \sum_{v=1}^{v=M} P_v \log_n P_v$$

where $P_v = n_v / N$
K = a constant

The principle of maximum entropy states in the present context that the candidate possessing the greatest value of S is most likely to be the object. The rationale for utilizing the principle is discussed in a recent review by Jaynes,[7] the associated methodology is termed the maximum entropy method, MEM. Because of the form of the entropy the chosen object will be flatter or smoother than other candidates, which is consistent with the true object being by definition noiseless.

Of course the number of candidates in any real situation is very large and thus it is by no means a simple task to find the object; nevertheless iterative cycles can be designed for this purpose. A particularly efficient and robust algorithm has been developed by Skilling and Bryan.[8] Their MEMSYS3 software package has been used for all examples of MEM deconvolution shown here.

Figures 9(*a*) and (*b*) illustrate the application of MEM to the Lorentzian doublet shown in Figure 3(*b*) and the IR spectrum in Figure 6(*a*). The results achieved are much superior to LDGA and no residual noise is evident in either case. Figure 10(*a*)–(*e*) compares the capacity of the various methods discussed in deconvoluting a slit function from the Raman spectrum of the stretching modes of CCl_4. The original spectrum in Figure 10(*a*) was recorded with a slit width of about 3 cm^{-1}.

The inverted second differential is shown in Figure 10(*b*) while deconvolution using a transfer function formed by Gaussian deconvolution with appropriate

[7] E. T. Jaynes, *Proc. IEEE*, 1982, **70**, 939.
[8] J. Skilling and R. K. Bryan, *Mon. Not. R. Astron. Soc.*, 1984, **211**, 111.

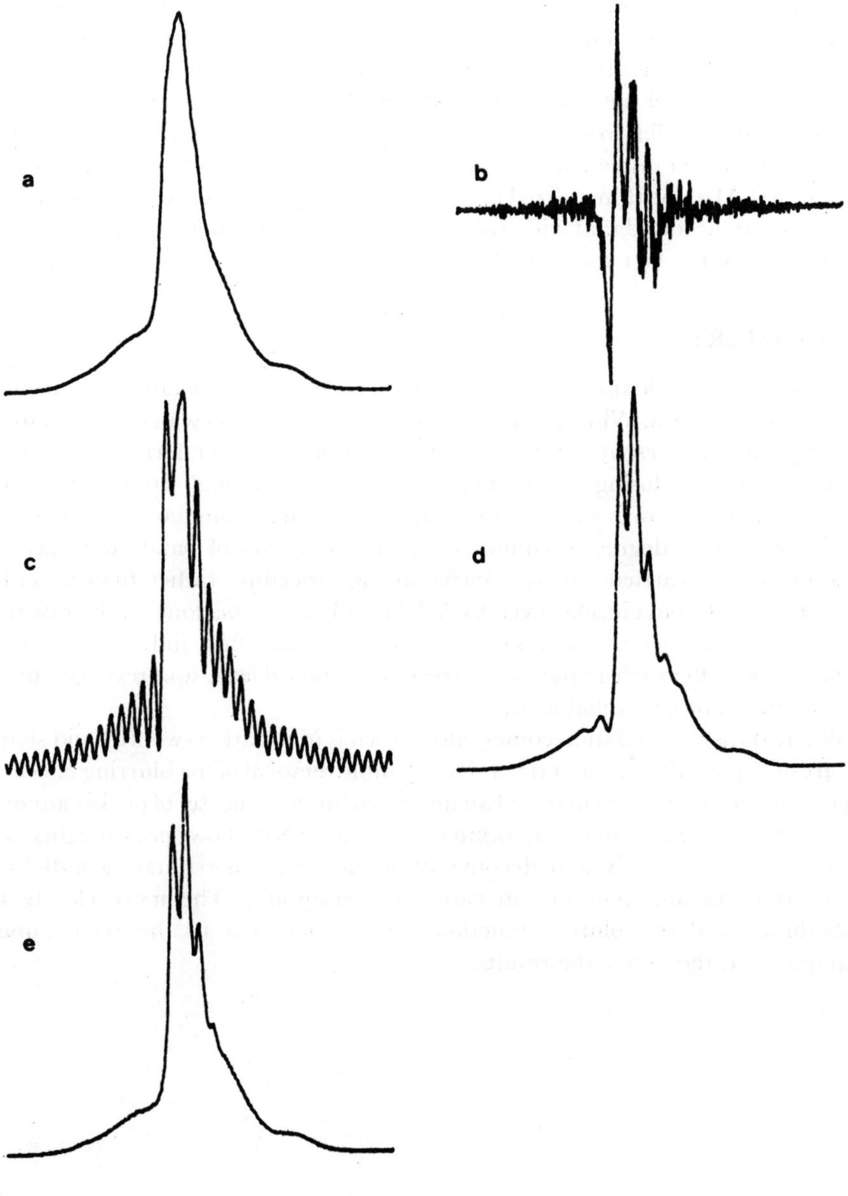

Figure 10 *Comparison of various methods used to enhance the resolution of the Raman spectrum of CCl_4 to reveal the isotopic splitting pattern*: (a) *original spectrum recorded at about* 3 cm^{-1} *slit width*; (b) *the inverted second differential of* (a); (c) *Gaussian deconvolution with Norton–Beer apodization of* (a); (d) *MEM deconvolution of* (a) *using the same Gaussian lineshape as for* (c); (e) *spectrum recorded with a slit width of the order of* 1 cm^{-1}

apodization produces Figure 10(*c*). MEM deconvolution using the same Gaussian blurring function gives Figure 10(*d*) which is quite similar to a spectrum recorded with much narrower slits ($\sim 1\ cm^{-1}$) shown in Figure 10(*e*).

Finally it should be noted that MEM deconvolution can, like ordinary deconvolution, suffer from incorrect specification of the blurring (line shape) function. However instead of generating negative side-lobes from over-deconvolution the MEM algorithm utilized here usually fails to converge to a final result (*i.e.* no object can be found that will give residuals as low as the noise variance). This is a reflection of the fact that the algorithm has been given the wrong task.

5 SUMMARY

Experimental considerations and circumstances often dictate the need to manipulate the data. While manipulation cannot add any new information to the spectrum it can present what is already there in a clearer form by removing unwanted and confusing effects such as noise and 'blurring'. However it is not always applicable; for instance if the shape of the background and the bands are known to a high degree of confidence then estimation of band strengths and positions is best carried out by a curve fitting procedure rather than by either smoothing or deconvolution (even by MEM). This must be done on the raw data as no benefit can accrue from, say, smoothing the data first, indeed because the residual noise after such a process is correlated, standard least-squares curve-fitting routines will not give reliable answers.

Where data manipulation comes into its own is in situations where band shapes *etc.* are only partially characterized. For example removal of the blurring effect of a spectrometer slit function to reveal an unknown 'underlying' set of peaks cannot be carried out by curve fitting. Smoothing to increase SNR allows peak heights to be measured more reliably and deconvolution can assist in estimating individual band strengths and positions in cases of overlapping. The more closely the smoothing or deconvolution function can be matched to the bands under manipulation the better the results.

CHAPTER 3

An Introduction to Fourier Theory and its Use in Spectroscopy

D. STEELE

Fourier transformation (FT) is now a common and familiar procedure in the spectroscopists laboratory. With the dedicated computer system attached to spectrometers the use of FT is generally a 'black box' procedure. Yet optimum useage of options, such as deconvolution procedures, does require at least a superficial understanding of what is involved. This chapter is directed to explaining the nature of FTs and of the important convolution theorem. The proofs are far from essential. We refer to any text on Fourier theory and that of Champeney[1] is particularly recommended as focusing on use by spectroscopists.

1 WAVES AND FOURIER TRANSFORMS

A disturbance of a quantity B (*e.g.* pressure or electric field) propagating in a direction x with a velocity c will have a value at a distance x and time t which can be written $A = B \times f(x,ct)$. For simplicity we are assuming that this wave is unattenuated, *i.e.*, it is not losing energy to the surroundings. For a simple cosine wave the amplitude can be written:

$$A = B\cos(2\pi(x/\lambda - ct/\lambda) + \varepsilon')$$

ε' is a phase angle depending on the defined origins of x and t. We can view this wave at a fixed time, in which case the amplitude as a function of x will repeat for intervals of x which are integral multiples of the wavelength λ. Alternatively at a given point x_0 in space we can observe the wave as a function of time. Again we see that B varies in a periodic manner, $\cos(2\pi ct/\lambda + \phi) = \cos(2\pi\nu t + \phi)$, where ϕ is a constant, the value of which depends on the amplitude at the chosen origin of time. ν is the frequency.

Any wave that is symmetric about some origin can be described as a sum, perhaps infinite in number, of cosine waves. To simplify the discussion we shall

[1] D. C. Champeney, 'Fourier Transforms and Their Physical Applications', Academic Press, London and New York, 1973.

assume a fixed value of time and concentrate on the changes in amplitude as a function of distance, written for a single cosine wave as

$$A = B\ \cos 2\pi\left(\frac{x}{\lambda} + \varepsilon\right) = \mathrm{B}\ \cos\ 2\pi(\bar{\nu}x + \varepsilon)$$

The reciprocal wavelength, or wavenumber, $1/\lambda$, is denoted $\bar{\nu}$. Thus a general symmetric wave can be described by

$$A(x) = \int_0^\infty B_i\ \cos 2\pi(\bar{\nu}_i x + \varepsilon)\ \mathrm{d}\nu_i \approx \Sigma_i\ B_i\ \cos 2\pi(\bar{\nu}_i x + \varepsilon)$$

We shall drop reference to the phase angle, although it is important in practical interferometry.

In practice we can never handle an infinite number of independent waves, and it is necessary to approximate by a direct sum. The approximation of taking the integral over all wavenumbers as a sum over a limited sample of wavenumbers is far from trivial. The effect will be analysed later. As will also be described later, one effect of a finite sampling interval in FT spectroscopy is to limit the highest frequency or wavenumber from which information can be derived.

If a cosine wave of a certain wavenumber is multiplied by a cosine wave of different wavenumber, and the result integrated over all possible values of x, then the result is zero.

$$\int_{-\infty}^{+\infty} \cos(2\pi\bar{\nu}_i x)\ \cos(2\pi\bar{\nu}_j x)\ \mathrm{d}x =$$

$$\tfrac{1}{2}\int_{-\infty}^{+\infty} [\cos 2\pi(\bar{\nu}_i - \bar{\nu}_j)x + \cos 2\pi(\bar{\nu}_i + \bar{\nu}_j)x]\ \mathrm{d}x$$

$$= 0,\ \bar{\nu}_i \neq \bar{\nu}_j$$
$$= 1,\ \bar{\nu}_i = \bar{\nu}_j$$

Using this theorem the distribution of amplitude as a function of wavenumber can be determined from the amplitude as a function of x.

$$B(\bar{\nu}_i) = \int_{-\infty}^{+\infty} A(x)\ \cos(\pi\bar{\nu}_i x)\ \mathrm{d}x$$

If the original data is asymmetric about the origin of x, then the cosine description is inadequate and a sine function has to be added. Both cosine and sine components are encompassed in the exponential form[2]

[2] Various definitions of the FT exist, differing in the presentation of the complex variables, or in the definition of the variables themselves.[1] It is common to define the transform as

$$B(y) = \frac{1}{2\pi} \oint A(x)\ \mathrm{e}^{ixy}\ \mathrm{d}x$$

This form results in some asymmetry in inverse and forward transformations and involves the radial frequency or wavenumber, $2\pi\nu$.

$$B(\bar{\nu}_i) = \int^{+\infty} A(x)\ \exp(i2\pi\bar{\nu}x)\ dx$$

since $\exp(i\theta) = \cos\theta + i\sin\theta$.

In Michelson interferometry the beam is split into two parts, and a varying path (phase difference) imparted between them before recombination. Let that phase difference be δ. On recombining the beams the total amplitude is obtained vectorially as $2A\cos(\delta/2)$. Now the power, P, of an electromagnetic wave is proportional to the square of the amplitude of the electric vector, so

$$P = 4K^2 \cos^2\left(\frac{\delta}{2}\right) = 2K^2(1 + \cos\delta)$$

$4K^2$ is the power when the waves are in phase, where K incorporates the proportionality factor. When the phase difference is 180° the power drops to zero.

Fourier transformation of the constant term must always be zero (the integral of a cosine or of a sine over a full cycle, and in the limit over infinity, is zero), and so we are left with the FT of the in-phase component $2K^2\cos\delta_i = K^2\cos(2\pi\bar{\nu}_i x)$ where i designates a chosen wavenumber. This generates a spectrum in the wavenumber domain.

In magnetic resonance the nuclear spin alignment in the applied magnetic field is randomized by external pulses and the realignment recorded as a function of time. Fourier transformation with the conjugate variable time gives the intensity as a function of frequency.

2 CONVOLUTION

The concept of convolution is best appreciated from an example. Consider a monochromatic wave of wavenumber $\bar{\nu}_0$ passing through a spectrometer with a symmetric arrangement of the optical elements about the grating. If the resolution is limited by the slit widths then the beam emerges from the exit slit, peaking at $\bar{\nu}_0$ but spread in a triangular distribution of intensity from $\bar{\nu}_0 - \Delta$ to $\bar{\nu}_0 + \Delta$. Δ is the spectral slit width or the spread of the beam in wavenumbers over the exit slit. The total intensity, or energy, is of course conserved. The beam is said to be *convoluted* with the slit function, or *apparatus function*. In Figure 1 the effect is shown of a triangular apparatus function on a Lorentzian band, as would occur for a slit width equal to the half height full width.

In spectroscopy it would be wonderful if we could somehow recover the true spectrum, removing all instrumental and environmental artefacts. Such a Utopian dream is unrealistic. Nevertheless, to a limited extent this can be achieved through the process of *deconvolution*. This means that we remove the distortions resulting from the convolution of the spectrum with the *estimated* distortion profile. There are a number of procedures, some based on iteration, and others on the Fourier theory. As the iterative methods are far more transparent we will begin with a discussion of these. The observed intensity $I(\bar{\nu})$ is related to the true $I^T(\bar{\nu})$ and to the instrument function $F(\bar{\nu}')$ by

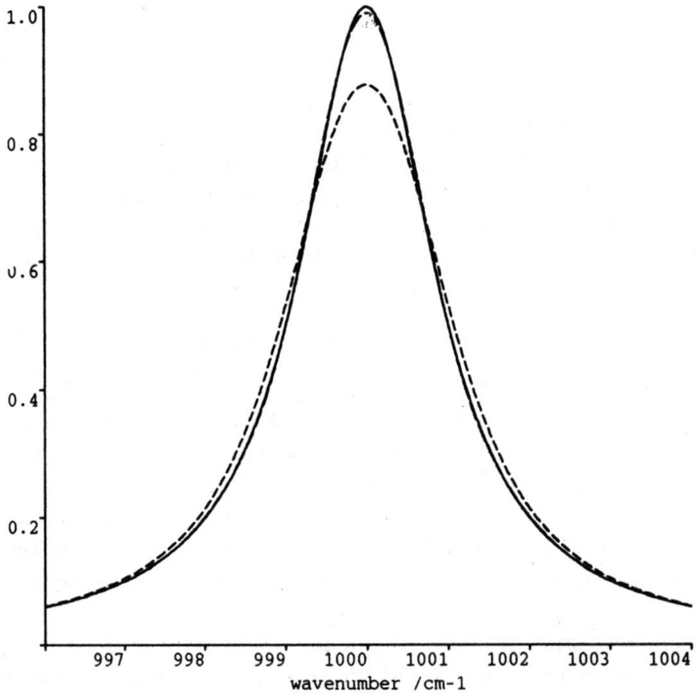

Figure 1 *Showing the effect of a triangular slit function on a Gaussian band as described in Table 1. ——— Original band shape; ---- distorted band shape*

$$I(\bar{\nu}_s)=\int_{-\infty}^{+\infty} I^{T}(\bar{\nu}_t)\ F(\bar{\nu}_s-\bar{\nu}_T)\ d\bar{\nu}_T$$

In the simplest iterative procedure due to Berger and van Cittert[3] one assumes initially that the observed intensities are the true and applies the instrument function to compute an estimated 'experimental', $I_{el}(\bar{\nu})$. The ratio of the observed spectrum to the computed, $I_t(\bar{\nu}_s)/I_{el}(\bar{\nu}_s)$, gives a first estimate of the correction factors by which each of the experimental intensities needs to be scaled to approximate the true values. Repetition using the new estimate for the 'true' spectrum and recomputing the 'experimental' quickly leads to a converged result. Such iterative procedures are valid only when a known apparatus function is available. Table 1 demonstrates the process for the earlier example of a Lorentzian convoluted with a triangular apparatus function. After four cycles the computed convoluted spectrum is in reasonable accord with the observed. Convergence becomes slow when the distortion is large compared with the original band intensity, *i.e.* when the resolution is significantly worse than the true band width.

In Fourier spectroscopy a much more versatile deconvolution procedure can be applied. Stated formally, the FT of the product of two functions is the FT of the

[3] H. C. Berger and P. H. van Cittert, *Z. Phys.*, 1932, **79**, 722: 1933, **81**, 428.

Table 1 *Showing the effect of convolution of a Lorentzian band,* I_{true} *with a triangular apparatus function of base width equal to twice the half band width of the Lorentzian. The band is centred at* 1 000 cm^{-1} *(only the high wavenumber half is shown) and the band width is* 1 cm^{-1}. *The apparatus function is equivalent to the use of a* 1 cm^{-1} *slit in a symmetric grating monochromator*

ν/cm^{-1}	I_{true}	I_{conv}	I_2	I_4
1000.0	1.000	0.878	0.974	0.990
1000.1	0.990	0.873	0.966	0.981
1000.2	0.962	0.859	0.944	0.956
1000.3	0.917	0.835	0.909	0.917
1000.4	0.862	0.803	0.863	0.866
1000.5	0.800	0.765	0.809	0.807
1000.6	0.735	0.722	0.749	0.743
1000.7	0.671	0.675	0.688	0.678
1000.8	0.610	0.627	0.626	0.615
1000.9	0.552	0.579	0.566	0.555
1001.0	0.500	0.532	0.511	0.499
1001.2	0.410	0.444	0.413	0.406
1001.4	0.338	0.368	0.337	0.335
1001.6	0.281	0.306	0.278	0.280
1001.8	0.236	0.256	0.233	0.237
1002.0	0.200	0.215	0.198	0.201

convolution of the two functions. Note that it is NOT the product of the FTs of the two functions treated independently. To clarify this consider the triangular apparatus function described earlier. Each element of the spectrum, $I(\bar{\nu}_i)$, is multiplied by $F(\bar{\nu}_j-\bar{\nu}_i)$ for $|\bar{\nu}_i-\bar{\nu}_j|\leq\Delta$, yielding an intensity contribution at the frequency (wavenumber) $\bar{\nu}_j$. The instrument function for a triangular distortion is

$$F(\bar{\nu}_i-\bar{\nu}_j)=\frac{1}{\Delta}\left(1-\frac{|\bar{\nu}_i-\bar{\nu}_j|}{\Delta}\right);\ |\bar{\nu}_i-\bar{\nu}_j|\leq\Delta$$
$$=0;\ |\bar{\nu}_i-\bar{\nu}_j|>\Delta$$

The resulting spectrum, $\phi(\bar{\nu})$, will be

$$\phi(\bar{\nu}_j)=\int_{-\infty}^{\infty}I(\bar{\nu}_i)\ F(\bar{\nu}_j-\bar{\nu}_i)\ d(\bar{\nu}_i)$$

The Fourier transformation of the experimental spectrum, $\phi(\bar{\nu})$, is the product of the FTs of $I(\bar{\nu})$ and of $F(\bar{\nu})$. Note that it is $F(\bar{\nu})$ and not $F(\bar{\nu}-\bar{\nu}_i)$ that is required. This is the convolution theorem. Provided that we can estimate the distortion, and hence $F(\nu)$, the 'true' spectrum can be obtained by first of all Fourier transforming the experimental data, $\phi(\bar{\nu})$, and then the instrument function, $F(\bar{\nu})$. If $\phi(\bar{\nu})$ is the intensity as a function of wavenumber, then t will represent the conjugate variable[4]

[4] The dimensions of the conjugate variable are the reciprocal of those of the variable itself. Thus the conjugate variable of frequency is time, and that of distance is the wavenumber. Taking the FT of a function of a variable produces a function of the conjugate variable.

of distance, namely x. Let the Fourier transformation of $I(\nu)$ and $F(\nu)$ be denoted $\mathcal{F}^-(I,t)$ and $\mathcal{F}^-(F,t)$, respectively. By the convolution theorum

$$\mathcal{F}^-(\phi,t)=\mathcal{F}^-(I,t)\times\mathcal{F}^-(F,t)$$

A simple general rule is that if two functions are multiplied in one space, then their FTs convolute. If they convolute in the first space, then their FTs multiply. Thus the unknown $\mathcal{F}^-(I,t)=\mathcal{F}^-(\phi,t)/\mathcal{F}^-(F,t)$. By carrying out an inverse Fourier transformation (substituting e^{ix} for e^{-ix}) *on* $\mathcal{F}^-(I,t)$ we derive the estimated 'true' spectrum $\phi(\nu)$; *i.e.*

$$\phi(\bar{\nu}_k)=\int_{-\infty}^{\infty}\mathcal{F}^-(\phi,t)\ e^{i2\pi(\nu_k-\nu_j)t}d(\nu_j)$$

Example 1 We can apply this theory to the earlier example. The Fourier transformation of a triangular function can be evaluated analytically since

$$\mathcal{F}\left[\left(\frac{1}{\Delta}\right)\left(1-\frac{(|\nu-\nu_c|)}{\Delta}\right),t\right]=\left[\frac{\sin(2\pi\Delta\frac{t}{2})^2}{2\pi\Delta\frac{t}{2}}\right]=\text{sin}c^2\left(2\pi\Delta\frac{t}{2}\right)$$

The Fourier transformation of the convoluted spectrum is divided by sinc^2, and the inverse FT generates the reconstituted spectrum. In this artificial data situation the true result is regenerated exactly provided that care is taken to avoid the singularities of the sin$c$ function, which occur at $t_1=2/\Delta$. However this is not a realistic example. There is a problem when sharp features occur in that the FT does not decay monotonically but shows strong oscillations a long way from the origin.[5] By cutting off the interferogram at some point $t_c<t_1$ a significant box function effect is introduced resulting in a band shape distortion (see Section 3). This is not then an option in attempting to correct for the apparatus function. It is better to remove the singularities by manipulation of the apparatus function transform.

Example 2 A more sophisticated application of the convolution theorem is in the separation of overlapping Raman bands due to isotopomers. The Raman spectrum of liquid chlorine consists of the contributions of $^{35}Cl_2$, $^{35}Cl^{37}Cl$ and $^{37}Cl_2$ in concentration ratios of 1 : 0.652 : 0.106. Let us designate these fractional ratios as x_m. Taking each isotopomer as having the same intrinsic band shape, and the isotopic shifts as $\Delta_{35,37}$ and $\Delta_{37,37}$, then the spectrum, $I(\nu)$, can be written as a function of the spectrum of $^{35}Cl_2$, $I^{35,35}(\nu)$.

$$I(\nu)=x_{35,35}I^{35,35}(\nu)+x_{35,37}I^{35,35}(\nu-\Delta_{35,37})+x_{37,37}I^{35,35}(\nu-\Delta_{37,37})$$
$$=\Sigma_i\ x_i\ I^{35,35}(\nu-\Delta_i)$$

[5] I. R. Hill, D. G. Owens, D. Steele and T. C. Jao, *J. Mol.. Struct.*, 1978, **43**, 117.

The sum i is over all of the isotopomer bands. Taking the FT about the band centre of the $^{35}Cl_2$ species, denoted ν_0, then

$$\mathscr{F}(I,t) = \int_0^\infty [\Sigma_j x_j I^{35,35}(\nu) e^{-i2\pi(\nu-\nu_0)t} e^{-i2\pi\Delta_j t}] d\nu$$

$$= [\Sigma_j x_j e^{i2\pi\Delta_j t}] \int_0^\infty I(\nu) e^{i2\pi(\nu-\nu_0)t} d\nu$$

The sum in the brackets [...] is the FT of the delta (δ) functions positioned at the centres of the three bands, each weighted by its fractional concentration. The integral is the FT of the spectrum of the $^{35}Cl_2$ species. By dividing the FT of the complete band complex by the FT as in the brackets, and then executing an inverse transform, the spectrum of a pure isotopic form is generated. In Figure 2 is shown the measured Raman spectrum of chlorine and the result of deconvoluting the effect of the isotopic species $^{35}Cl^{37}Cl$ and $^{37}Cl_2$.[6]

3 RESOLUTION

It is impractical to record a spectrum over an infinite time interval, or over an infinite range of displacement in a Michelson interferometer. The consequence of a finite range is to limit the resolution in the frequency/wavenumber domain. Precise definition of the resolution of a system requires care when specifying that two objects are actually resolved. Is it sufficient, for example, that a minimum should occur, however small, between the positions of the bands – or must there be a specific dip defined as a portion of the peak heights? In any case it is usual to specify when discussing resolving power that the bands should be of equal height. In spectroscopy it seems logical to require that for two component frequencies to be resolved a path difference of one wavelength must exist for extreme rays in the optical system. In an interferometer this necessitates that the maximum path difference recorded, d, should be equal to, say, $(n+1)\lambda_1$ and $n\lambda_2$ where λ_1 and λ_2 are the two wavelengths which can just be resolved and n is an integer. Let the reciprocal wavelengths, or wavenumbers, be $\tilde{\nu}_1$ and $\tilde{\nu}_2$. Then

$$\frac{(n+1)}{\tilde{\nu}_1} = d = \frac{n}{\tilde{\nu}_2}$$

or by eliminating d,

$$\tilde{\nu}_1 - \tilde{\nu}_2 = \frac{1}{d}.$$

More rigorous approaches to this problem through Fourier theory can be made in various ways. I shall not endeavour to labour this issue, since the result given above is so well known. Nevertheless, it is instructive just to indicate, rather than prove two approaches. In the first approach it can be shown (see *e.g.* ref. 1) that if

[6] E. B. Gill and D. Steele, *Mol. Phys.*, 1977, **34**, 231.

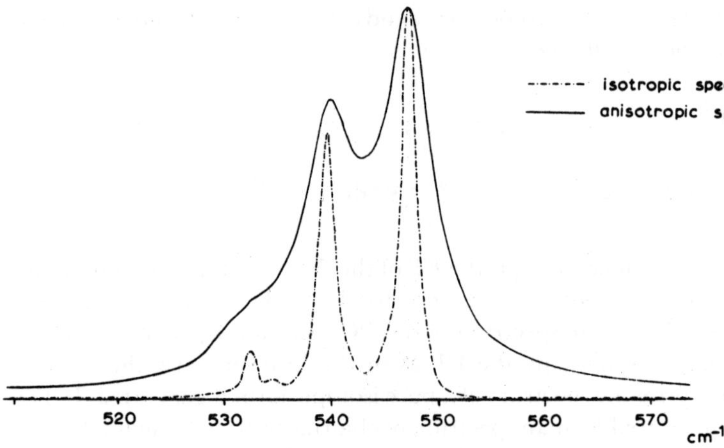

(*a*)

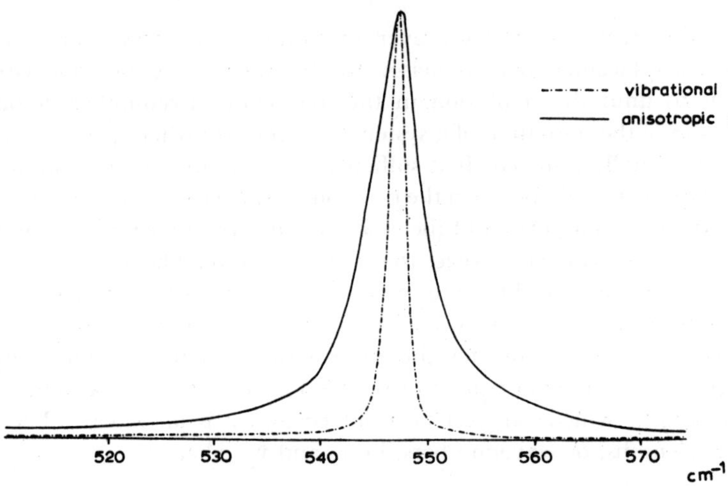

(*b*)

Figure 2 (a) *Isotropic and anisotropic spectra of liquid chlorine at* 178 K. *(Arbitrary intensity scale.)* (b) $^{35}Cl_2$ *spectra at* 178 K, *calculated by correcting the functions* G_{2V} *and* $G_{2V}G_{2R}$ *for isotopic and hot-band broadening, and then re-transforming the resultant functions*

we have two functions which are the FTs of one another, then the second moments of these functions, $|f(x)|^2 = \Delta x$ and $|F(y)|^2 = \Delta y$, are related by $\Delta x \Delta y \geq 1/4\pi$. In these definitions we assume that F(x) and F(y) are centred around the origin.

Note the inverse behaviour. As the function F(y) narrows so f(x) broadens. Assuming a wide distribution of information in one domain, say in F(y), representing a rich interferogram in time or distance, then the widths in the frequency or wavenumber domains will be inversely proportional to the period of

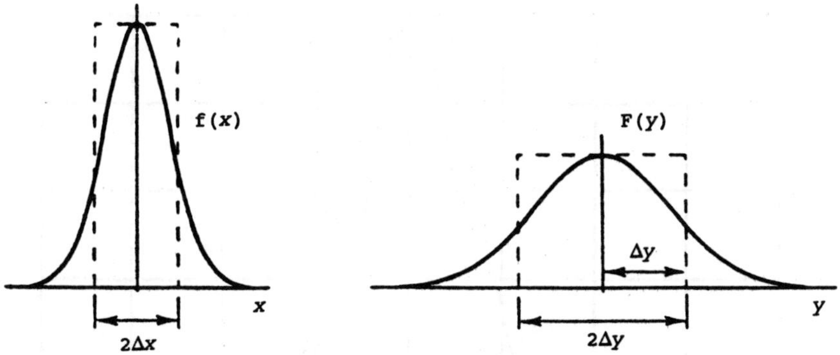

Figure 3

measurement in F(y). Thus the resolution is seen to be inversely proportional to d. Of course a pure harmonic frequency wave produces a continuous transform, thus perfectly satisfying this requirement. The 4π factor relates to the half band widths, and clearly the specification of resolution requires further work. Two gaussian bands

$$I(\nu)=\exp[-a(\nu-\nu_0)^2]$$

of equal magnitude and of half band widths $d_{1/2}$ will show a 7.3% dip between them when they are a distance apart of $\delta_{1/2}$.[7] If an interferogram is recorded of such a two band spectrum and the length of the interferogram is $1/\delta_{1/2}$ then the dip will be reduced to 5.0%. Thus we see that the earlier requirement of a full wavelength separation for extreme rays, which led to the resolution being $1/d$, is a perfectly satisfactory one.[8]

A similar result is obtained easily *via* the convolution theorem. If an interferogram is truncated after a period t_{max}, then this is equivalent to its multiplication by a box function ⊓ of base length $2t_{max}$. ⊓ takes the value of 1 within the limits of the measurement interval, say t_{max} to zero.

$$\phi'=\phi\times\sqcap$$

Taking the FT and using the convolution theorum

$$F(\phi')=F(\phi)\otimes F(\sqcap)$$

where ⊗ denotes a convolution. But F(ϕ) is the spectrum as it would have

[7] At the points $\nu\pm\delta_{\frac{1}{2}}$, $I\,(\nu_0\pm\delta_{\frac{1}{2}})=\dfrac{I\nu_0}{2}=\exp\,(-\,a\delta_{\frac{1}{2}}^2)$

[8] Note that taking the fundamental quantum relation between energy and frequency, namely Planck's Law

$$E=h\nu$$

then the second moment relation $\Delta\nu\Delta t\geq 1/(4\pi)$ translates into Heisenberg's Uncertainty Principle.

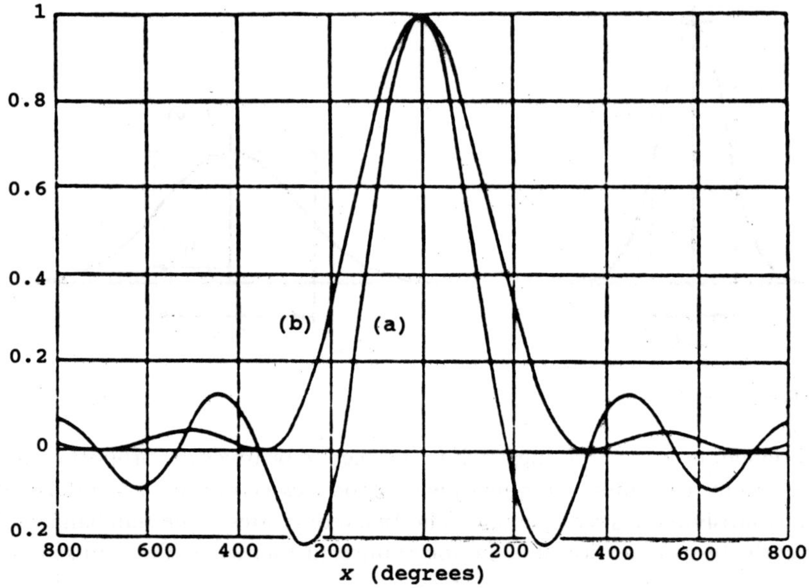

Figure 4 *(a) Function* $\frac{\sin x}{x}$, *(b) function* $\left(\frac{\sin x/2}{x/2}\right)^2$

appeared without the effect of truncation and thus the observed spectrum will be this convoluted with the FT of the box function. The box function can be written (assuming the variable $0 \leq t \geq t_{max}$) as

$$1, \ |t| \leq t_{max}$$

$$0, \ |t| \geq t_{max}.$$

Its FT is $\sin(2\pi t_{max}\nu)/(2\pi\, t_{max}\nu) \equiv \mathrm{sin}c(2\pi t_{max}\nu)$. The sin*c* function (see Figure 4) causes broadening and also side-lobes. The magnitudes are not unappreciable. In an interferometer ($t \rightarrow$path difference and $\nu \rightarrow$wavenumbers) for a mirror movement of 5 cm, leading to a path difference x_{max} of 10 cm, the width as defined by the first abcissa crossing of the sin*c* function is $2\pi/5\ \mathrm{cm}^{-1}$. It is for this reason that the interferogram is apodized, *i.e.*, the data is subjected to additional weighting, for example by a cosine function $\cos(\pi x/2x_{max})$. This eliminates the discontinuity at $x = x_{max}$ and thereby eliminates the negative lobes in the spectrum. For further discussion of this the reader is referred to texts on interferometry (*e.g.* ref. 9).

Naturally if a real band has a width far greater than $1/x_{max}$, then increasing x_{max} is not going to improve the resolution. This sounds like a trite statement and yet many errors and a lot of wasted effort have gone in the past into attempts to beat

[9] A. E. Martin, 'Infrared Interferometric Spectrometers', Vol. 8 of 'Vibrational Spectra and Structure', ed. J. R. Durig, Elsevier Press, Amsterdam, 1980.

nature in this. The most common computer based procedure to sharpen bands is based on deconvolution procedures. Consider the spectrum $I(\nu)$ convoluted with an apparatus function $\mathrm{A}(\nu)$. The measured interferogram $\mathrm{f}(x)$ is given by

$$\mathrm{f}(x) = \mathscr{F}^{-}[I(\nu) \otimes \mathrm{A}(\nu)]$$

$$\text{thus } I(\nu) = \frac{\mathscr{F}^{+}\ [\mathrm{f}(x)]}{\mathrm{A}(\nu)}$$

where $\mathscr{F}^{+}$ represents the inverse transform. That is, as we have seen earlier, we can deconvolute the observed spectrum to remove the apparatus function, provided that we know the latter. Suppose that we now manipulate $\mathrm{A}(\nu)$ to yield a broader more slowly decaying function in the conjugate domain, x. The effect by our earlier arguments will be to narrow the apparent spectral features.

From a realistic viewpoint, if an instrument cannot resolve two features because it is beyond the optical étendue of the interferometer, or because it is beyond the grating resolution, procedures of forced narrowing by deconvolution must be questionable. To obtain an insight into what occurs on forcing band narrowing, consider what happens when we take the differential of the spectrum. The FT of the nth differential of a function, $\mathrm{f}(x)$, is $(\pm iy)^n$ times the FT of the function itself, where y is the conjugate variable to x; *i.e.*

$$(-iy)^n\ \mathscr{F}^{+}\ (y) \leftrightarrow \frac{\mathrm{d}^n\ \mathrm{f}(x)}{\mathrm{d}x^n}$$

We see that a simple doctoring of the interferogram produces a differential spectrum, and of course the differential spectra are narrower. A second differential spectrum does yield a clearer picture of the number of components present in the spectrum, but at a significant cost. In the even differentials side-lobes are created and intensity data is lost.* To emphasize the latter consider the differential of a small infinitely narrow noise spike. The first differential is infinite! In practice it is often more useful to use truncated FTs to remove noise, since by curtailing the interferogram the higher frequency signals are smoothed out. Note that they are eliminated, but their average may form a smoother displaced background.

4 EFFECTS OF FINITE SAMPLING INTERVALS

By sampling at regular intervals of distance or time the spectrum is in effect multiplied at by a regular comb of unit delta functions. Since we have a multiplication in the sampling domain, then in the transform domain the spectrum is convoluted with the FT of the comb function. The FT of a unit comb function is itself a comb delta function. Assuming unit weighting to the comb and a sampling interval of Δx, then the FT is a unit comb with sample spacing of $1/(\Delta x)$. To appreciate the impact of this, note that if sampling occurs every 10^{-3} cm then

* See Chapter 2, Section 3.2.

in the wavenumber domain every spectral element recurs every $10^3\,cm^{-1}$. Any component in the spectrum with a wavenumber greater than $1000\,cm^{-1}$ would then be overlaid. It is essential therefore that all energy at wavenumbers greater than the reciprocal of the sampling interval be filtered out.

These are just a sample of the numerous applications of the convolution theorem in spectroscopy, but by examining these well known problems it is hoped that readers will feel less intimidated in developing their own applications.

CHAPTER 4

Spectroscopic Databases

A. N. DAVIES AND P. S. McINTYRE

1 INTRODUCTION

Spectroscopic databases are playing an ever increasing role in the life of the spectroscopist. The quality and sophistication of the database systems available is very much dependant on the type of spectroscopy, but the general development direction is now becoming clear. Databases with integrated expert systems containing information from different types of spectroscopies should be common tools for the analyst of the 21st century. This is still only a poor substitute for the service which is really required by spectroscopists where the software systems should be capable of analysing a given spectrum or set of spectra and producing a correct identification of the compound or mixture being analysed without requiring time consuming referrals to reference databases.

This chapter will present an overview of the different spectroscopic techniques that use databases, a review of the databases currently available and a short comparison of the type of search algorithms used.

2 HISTORICAL PERSPECTIVE

For many years spectroscopic data collections appeared in printed form. The typical content of these collections was the spectral curves, plotted on various axes, peak tables and a large amount of information on spectroscopic constants. Most practising spectroscopists have carried out the gruelling task of searching through such books trying to establish to what class of compounds a particularly puzzling unknown substance belongs.

The introduction of computers into the spectroscopic work-place should have made such searching obsolete. In certain types of spectroscopy steadily growing database collections are available. These databases can be installed on the spectrometer control computers or in third party software packages for data analysis away from the measuring instrument. The largest collections of spectra are available for organic mass spectrometry but even here only a tiny proportion on the total known compounds are represented. Unfortunately, even to date, the majority of spectra measured are published in printed form and are not made available for digital databases, even though they are all obtained in the laboratory in digital form.

For years the expansion of spectral databases has been held back by the relatively large amount of computer storage required and the poor data processing speed of the available computers. Many publications have appeared in the literature covering different methods of speeding up database searching.[1–4] For some spectroscopies one method has been to store the data in a binary format simply indicating the presence or lack of peaks at a particular x-axis position, but this method enormously reduces the amount of information stored. It should be noted that the speed of change in hardware, CPU speed and storage media has not been matched by software writers. In other words the hardware is not used to its full capacity, shortcuts are still used in storage to minimize space perhaps to the detriment of the success of the search.

3 DATABASES, EXPERT SYSTEMS AND ARTIFICIAL INTELLIGENCE

All too often computer systems containing extensive stored information are labelled 'expert systems' or even worse described as exhibiting 'artificial intelligence'. The mystique built up around these definitions has often led to misuse and to heated discussions as to their real meaning.

A database contains a number of data sets which can be automatically searched through by a computer program. The search should yield a sub-group of the data collection which corresponds to that desired by the user. The important features of the selected data should be made available to the user either on screen or in printed form.

An expert system consists of a knowledge base, generally a database of reference facts, and a list of rules or guidelines generally derived from the knowledge base with the intention of producing a system which, using these rules, can create conclusions about any new data set presented. This can roughly be described as if fact X is true then conclusion Y must be drawn, or, on a more spectroscopic example, if a peak is found between 1700 and 1750 cm^{-1} then the unknown compound contains a carbonyl group.

Artificial intelligence is a property, and an expert system is sometimes mistakenly described as having artificial intelligence. For a spectroscopist the key to understanding the difference lies in the inability of an expert system to draw sensible conclusions about data from areas of chemistry not contained in the reference data set. Two recent review articles by Wendy Warr on computer-assisted structure elucidation deal with library search and spectral data collections as well as indirect database approaches and established systems and are well worth reading.[5]

[1] M. L. Olson, *SPIE*, 1981, **289**, 236.
[2] G. Hangac, R. C. Wiebolt, R. B. Lam and T. L. Isenhour, *Appl. Spectrosc.*, 1982, **36**(1), 40.
[3] M. F. Delaney, J. R. Hallowell and F. W. Warren, *J. Chem. Inf. Comput. Sci.*, 1985, **25**, 27.
[4] S. Kawata, T. Noda and S. Minami, *Appl. Spectrosc.*, 1987, **41**(7), 1176.
[5] W. A. Warr, *Anal. Chem.*, 1993, **65**(23), 1045A; 1993, **65**(24), 1087A.

4 DATABASE QUALITY

The requirements needing to be met by a top quality database can radically differ between different types of spectroscopy, and the use to which the data is going to be put. Anyone requiring a simple answer to the question 'what is a good database?' must first clarify what they are going to do with the system, but also to keep in mind how easy it is to use.

A useful starting point is to divide up the database system into its two constituent parts. A database consists primarily of a data library and a piece of software designed to search in this data.

4.1 The Data Library

The data library should be built from spectra of pure substances, whose purity has been ascertained by, for example, chromatography, or by at least one other secondary verification (melting point, boiling point *etc.*). *NB* Reference data as defined here should not be confused with archive data which is a simple record of the spectra having been recorded and the purity or lack of the substance which was measured is then irrelevant!

For reference data the molecular structure should be stored as a connection table and be searchable. The usefulness of three-dimensional structure coding will only be relevant in spectroscopies sensitive to this aspect of molecular structure. Isotopic information and isomerism can be useful if known, but the information should never be guessed as it may be of critical interest in the future for a user of the reference data. A general rule should be 'if the parameter is not known – leave it out!'.

Data manipulation should not have been carried out on the data in the reference collection. If any data manipulation has been carried out this must be documented and be made obvious to the user of the database (baseline correction, smoothing *etc.*) As much instrument parameter information should be made available to the user as is practicable with regard to the intensities measured and the spectral resolution *etc.* The accompanying factual information should be as complete as possible covering Chemical Abstracts Service (CAS) registry numbers, for substance, CAS name, origin and copyright owner of the spectral data, as well as a contact address for anyone requiring more information about the particular reference data set. Information on the sample handling should be available both for the sample preparation before the spectra was measured and for the general sample collection. Furthermore, cross referencing to relevant information held elsewhere should also be as extensive as possible. As a rough guide, any information held on a spectrum should be placed in the database even if the immediate relevance is not clear. Some future user may well thank you for including the actual room temperature! (Readers in Southern California please note!)

4.2 The Search Software

The algorithms used to search through the reference spectral data are as important as the data itself. Many different authors have researched the strengths and

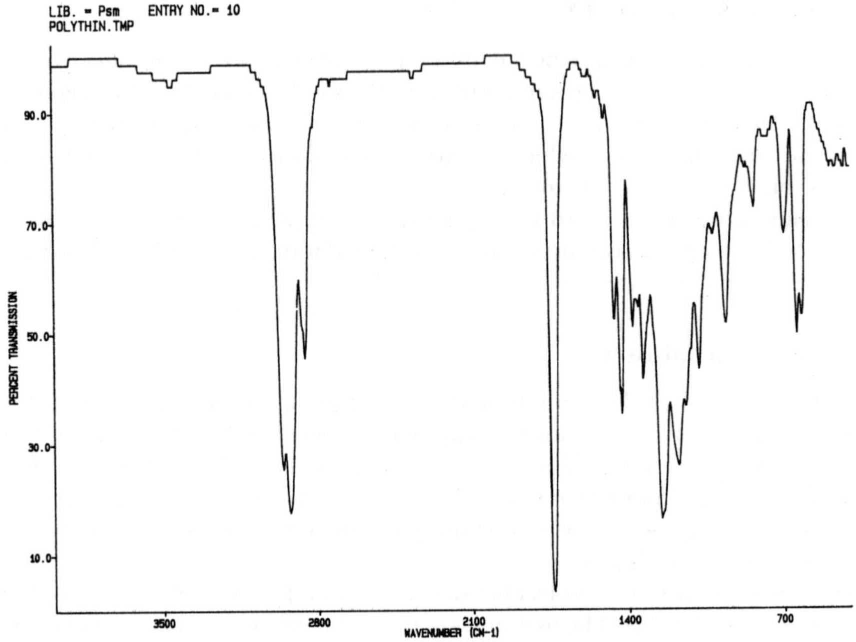

Figure 1 *Transmission spectrum*

weaknesses of different search strategies (*e.g.*, refs. 6 and 7).

The search software can also be technique dependant, so we shall quickly introduce the most important of these, then compare three of the commonest search algorithms.

The four commonest techniques are ultra-violet–visible (UV–VIS), mid-infrared (IR), nuclear magnetic resonance spectroscopy (NMR) and mass spectrometry (MS). UV–VIS spectroscopy gives broad absorption bands and few of them, so low resolution spectra are acceptable for database building. However for this reason the technique is not important for most structure determination problems, but has been used in limited areas.

IR spectroscopy has sharper bands, many of them, and the spectrum has structure dependent information; for example all chemists would be able to spot a carbonyl compound using this technique, without the help of a database! The resolution of the stored spectra in databases should therefore be higher than for UV–VIS. IR databases are commonly used to identify unknowns, usually against a small list, or identify functional groups. IR poses large problems for commercial library builders. The problem lies in the strengths of the technique viewed from other standpoints. It can be used for solids (here commonly as ground suspensions in liquid paraffin or dispersed in potassium bromide), liquids (in solution or as neat liquids) and gases (what about pressure and its effect on the spectrum?). Spectra

[6] P. B. Harrington and T. L. Isenhour, *Anal. Chem. Acta*, 1987, **197**, 105.
[7] J. T. Clerc, E. Pretsch and M. Zürcher, *Mikrochim. Acta*, 1986, **II**, 217.

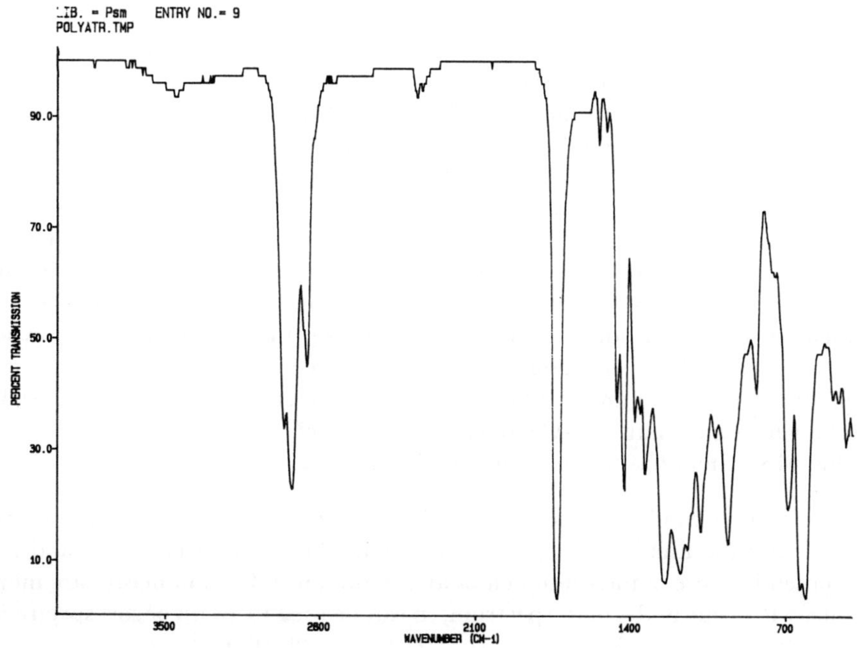

Figure 2 *Reflectance spectrum*

obtained for the same compound by these different methods give very different spectra. The human eye and brain can cope with this, but search software is going to get confused! Compare Figures 1 and 2, they are of the same polymer, one as a transmission spectrum the other as a reflectance spectrum. To our eyes they are the same, but not to a search algorithm; the intensities are very different below 1400 cm^{-1}.

NMR spectroscopy has very sharp bands for spin $\frac{1}{2}$ nuclei, can often have hundreds of them and contains much more information than IR or UV–VIS. So, the resolution of stored spectra must again be high, but here, unlike IR spectroscopy, there are often large areas within the 'standard spectral region', of 0–10 p.p.m. for ^1H spectra or 0–250 p.p.m. for ^{13}C spectra, that have no absorption bands, wasting space on the storage medium. NMR is also a technique that has taken advantage of the computer revolution, many two-dimensional techniques having been introduced. This complicates database use and has led to extensive discussion about storage formats for databases and methods of transfer between instruments or databases.

MS is perhaps the easiest of the four common methods for database building. So-called low resolution spectra (ones in which abundance of the ions is given a whole mass number) simply have a peak above a threshold value or not, and the peak has a width of one data point (very unlike UV–VIS or IR). Gas chromatography–MS (GCMS) is routinely used to identify the components of mixtures; these may be entirely unknown so a large database is used or the laboratory may be one specializing in drugs of abuse, so here a small library can be used (only those compounds defined as drugs of abuse in law need be in the library), usually with very good results.

So how do search algorithms work? There are as many solutions to this as there are commercial systems and lots of others that have not reached the market; ref. 7 compares 41 different systems for IR alone. However, they do fall into types, and we shall look at two IR ecamples and one MS example.

As mentioned above mass spectra are easier to deal with than infrared spectra. Certain ions are more stable than others, *i.e.* have a longer lifetime in the instrument, so reach the detection system in large amounts. This is as a direct result of their structure so is ideal for database builders – the commonest ions have a structural significance, so should be kept and accentuated. Conversely, small peaks are insignificant and can be discarded, unlike IR, where less intense peaks can be very significant (C=C stretch, C≡N stretch *etc.*).

One MS system has the following attributes:

1 The system condenses spectra, *i.e.* allows them less space. This is done by selecting the number of peaks to be stored, this varies between 15 and 26, chosen because of their uniqueness and abundance. Certain masses are more likely to occur in the mass spectrum, so can be used to group or sort spectra in the database, *e.g.* 91 ($C_7H_7^+$), 43 (CH_3CO^+), 28 ($C_2H_4^+$) *etc.*
2 A comparison is then made of the most significant peak in the unknown spectrum against the library. This is then repeated for the next most significant peak, *etc.* This is fast, the stored spectra are small, so libraries can be large (>100 000). They tend to work better than IR libraries, because of the sampling problems outlined above.

The search software works as follows:

- A prefilter selects the most significant peaks in the unknown
- Spectra are then selected from the library
- Base peaks in the reference are compared with the unknown
- Missing peaks are flagged
- The reference spectrum is scaled
- A confidence value is assigned
- These confidence values are sorted and the list printed to the computer screen.

Typically, IR libraries are much smaller than MS ones. This historically reflects the storage space on $5\frac{1}{4}''$ floppy disks, since most of these systems existed before hard disks were available. It also reflects the large amount of space required for IR over MS spectra. Here, typically, organic compound libraries are 2 000 spectra, polymer libraries 600 and gas phase libraries 3 000. Note the subdivisions: IR spectroscopists use specialist libraries if possible, as this speeds up the search, and makes the identification more reliable. Note that the gas phase libraries are larger; this is a reflection of the timing of their production, to aid analyses in GC–FTIR when hard disks were already in general use.

A typical commercial IR system has the following attributes and method of searching:

- Users can create their own libraries. This has obvious advantages.
- Unknown spectra can be imported in many different formats, usually in JCAMP–DX; many different formats are recognized.
- Spectra can be searched using a peak table (frequency of the main peaks in the spectrum, generated by the system), or on the whole spectrum, or a combination of both.
- Chemical structures are also stored, so can be output to reports; physical data are stored, and can be searched.

The search software works as follows:

1 The unknown spectrum is scaled in the same way as the library; the baseline can be corrected for scattering, a common problem in IR of solids.
2 The unknown is then subtracted from every entry in the library. This results in a difference spectrum, containing positive and negative peaks.
3 The hit quality is then the sum of the squares of these differences. This is inevitably a slow process!

The second type of IR search algorithm claims elements of an 'expert system', and has the following attributes:

- It can recognize functional groups
- It can handle mixtures
- Searches on functional group information and also on a peak table, and it can also search on chemical information
- It produces five different hit lists, to cater for the mixture option
- Users can create own libraries
- Cannot import JCAMP spectra, so is restricted to one manufacturer
- The system can however export JCAMP spectra, so the system above can also be used.

The system works as follows:

1 The unknown spectrum is converted into a peak table, using a predefined or user input threshold value for absorbance.
2 This is scanned to produce a list of possible structural units obtained by comparison with very many stored possibilities. The library spectra also have these structural units stored.
3 The stored units are compared for the two spectra. Peaks in the spectra are also compared and finally any chemical information can be compared.

This is therefore a complex area, especially in IR, where sampling method has a large input into the final spectrum. Modern Fourier transform instruments have good calibration of frequency so this has improved. Older dispersive spectra can have 5% errors in frequency calibration, again leading to problems in database construction.

4.3 User Friendliness

One point which is ignored all too often is that any system must be capable of being understood by laboratory technicians and professors alike, people not trained in computing. For users to be able to use the database comfortably and successfully the interface to the user must be relatively simple, with built-in help and guidance of relevance to the beginner as well as to the expert. These help texts must be in normal language and not laden with computer jargon. If these aspects are missing in a database system it will either not be used, or what is probably worse, used incorrectly. The dangers arising from the misuse of a spectroscopic database are false identification of chemical substances, failure to identify risks properly, impurities in chemical production *etc.* This could then lead to increased health and safety problems, loss of production, failure to recognize problems at a sufficiently early stage, and most likely, to dramatically increased costs through decision making based on incorrect data.

The best way to maintain the quality of a reference collection is to ensure it is simple enough to be used regularly by as wide a range of spectroscopists as possible. If feedback from the users is then encouraged, errors can be weeded out rapidly (*e.g.* ref. 8).

Let us now look at the type of libraries that are available for the different spectroscopic techniques.

5 REFERENCE COLLECTIONS

5.1 Mass Spectral Databases

Mass spectrometry is the area of most success as far as databases are concerned. Three major collections are widely available, and what is more important, widely used. Their history and development have been well documented elsewhere.[9]

The British Government initiated a data collection project at the Atomic Weapons Research Establishment (AWRE), Aldermaston, which is known as the Mass Spectral Data Centre (MSDC). The MSDC reached *ca.* 25 000 complete spectra with *ca.* 70 000 available as an eight peak database.

The US National Institute of Health (NIH) computerized the collection and later with the US Environmental Protection Agency (USEPA), the then National Bureau of Standards (NBS), the Food and Drug Administration (FDA) and the MSDC enlarged this database and introduced quality control of the data entering the collection. None of the spectra in this database are isotope labelled but an advantage is that only one spectrum per compound is entered into the database. This NBS database can be searched online through a number of hosts.[10,11]

The third major database was known as the Mass Spectrometry Registry which was started by Stenhagen and Abrahamsson in Sweden and continued by McLafferty, published by Wiley and Sons. However, multiple spectra per

[8] G. R. B. Webster, L. P. Sarna, D. F. Kenny, W. D. Buchannon and J. B. Westmore, *J. Agric. Food Chem.*, 1993, **41**(11), 1993.

[9] S. R. Heller, 'Mass Spectrometry Databases and Search Systems', ch. 6 of 'Computer Supported Spectroscopic Databases', ed. J. Zupan, Ellis Horwood, Chichester, UK, 1986.

[10] Telesystems-Questel, 83–85 boulevard Vincent Auriol, 75013 Paris, France.

[11] Chemical Information System Inc., a subsidiary of Fein–Marquart Associates, 7215 York Rd, Baltimore, MD 21212, USA.

compound were allowed if the entries were significantly different from one another and radio labelled spectra were also included. Chemical Abstracts numbers are available for substances in the database as well as chemical structure information. The Quality Index for this database is slightly different to that used by the NBS database. The collection is widely available and is also available on CD-ROM. In 1991 the database was reported to have been substantially improved and enlarged to 139 859 spectra records for 118 144 compounds which include CAS registry numbers.[12] It can be found in certain third party PC software packages.[13,14]

The NBS and Wiley databases are combined into the Wiley–NBS Registry of Mass Spectral Data containing over 150 000 spectra of 112 272 compounds with structures provided for 89 903.[15] The computer version is slightly smaller with about 140 000 spectra but from 118 144 compounds.[16]

The Wiley–NBS database is also available online combined with a second database of *ca.* 10 000 spectra from Henneberg and running under the Henneberg software.[17]

A report in 1988 discussed the Japanese Information Centre of Science and Technology (JICST) mass spectral database. This database, which began in 1987 and was restarted in January 1988 as part of JICST Factual database system, contains, besides MS, thermophysical and thermochemical properties, chemical substance names and structures.[18]

Many small mass spectral databases of speciality chemicals have been reported, some examples of which will be detailed below.

A GCMS database of more than 1 600 drugs supplemented by an in-house library has been running for several years successfully analysing urine and serum samples and was developed to run automatically.[19]

A database of phencyclidine [1-(1-phenylcyclohexyl)piperidine or PCP] and 128 related compounds studied by GCMS is available on floppy disk.[20] A GCMS jet fuel database (complex hydrocarbon mixtures), retention times and mass spectra has been recorded.[21] Other GCMS based information systems contain data, for example, on organic phosphorous and chlorine containing pesticides, low boiling point organic chlorine containing solvents, and low boiling point aromatic compounds developed for use in identifying unknown pollutants in the

[12] F. W. McLafferty, D. B. Stauffer, A. B. Twiss-Brooks and S. Y. Loh, *J. Am. Soc. Mass Spectrom.*, 1991, **2**(5), 432.

[13] E. G. Smith and P. A. Baker, Chemical Information Management Inc., Cherry Hill, New Jersey, USA, 1975.

[14] 'The CD-ROM Directory', TFPL Publishing, London, UK, 2nd edn, 1988. For IBM PC or compatible with 640K RAM. Use of Phillips, Sony, or Hitachi CD-ROM readers recommended.

[15] F. W. McLafferty and D. B. Stauffer, 'The Wiley–NBS Registry of Mass Spectral Data', John Wiley and Sons Ltd, Chichester, UK, 1989.

[16] F. W. McLafferty and D. B. Stauffer, 'The Wiley–NBS Registry of Mass Spectral Data', John Wiley and Sons Inc., Sci/Tech Division, New York, USA, 1989.

[17] D. Henneberg and B. Weimann, *Österr. Chem. Z.*, 1988, **89**(11), 328.

[18] K. Okano and A. Abe, *Joho Kanri*, 1988, **31**(1), 56.

[19] N. B. Smith, *J. Anal. Toxicol.*, 1994, **18**(1), 16.

[20] D. J. Gole, J. L. Pirat, J. M. Kamenka, A. Kalir and E. F. Domino, 'Sigma Phencyclidine-like Compd. Mol. Probes. Biol.' (Proc. US–Fr Sponsored Int. Seminar), ed. E. F. Domino and J. Kamenka, NPP Books, Ann Arbor, Mich. USA, 2nd, meeting date 1987, 78–79.

[21] E. M. Steward and E. W. Pitzer, *J. Chromatogr. Sci.*, 1988, **26**(5), 218.

environment[22] or chlorination by-products in drinking water.[23] A positive chemical ionization mass spectra database of trichothecenes from quadrupole MS measurements has been reported using the software packages PBMULT and PEAKS to control the system.[24]

Other databases cover GCMS identification of solvent mixtures, light oils and metabolites in human urine,[25] pyrolysis MS of polymers,[26,27] composite spectra generated from more than 1 500 commonly available adhesives measured during pyrolysis-capillary-column GCMS used to study improvised explosives.[28] A GCMS database of organic pollutants in sewage sludge using chemical ionization methods has been reported,[29] as have collections of low resolution mass spectra of methyl esters of *N*-decanoyldipeptides used for determining mixtures of dipeptide derivatives,[30] fatty acid methyl esters,[31] N-containing aromatics,[32] low resolution steroid mass spectra[33] and pyrolysis GCMS spectra of lignocelluloses.[34] A novel search system with a new mass spectra database of 2900 environmentally relevant samples called NIES-MSLS has been developed at the National Institute for Environmental Studies in Tsukuba, Japan.[35]

GCMS has also led to the production of specific databases following biodegradation of oil[36] and mixed data databases where the Kovats retention indexes are also stored.[37,38] This combined use of the retention time data and the spectra has been utilized in an automated method for the identification of fire debris samples.[39]

A useful database for secondary ion mass spectroscopists could be a database of sample sputtering and beam size characteristics from a gallium ion microprobe, which reportedly can be used to determine settings required for successful secondary ion mass spectrometry.[40]

The Royal Society of Chemistry and Mass Spectrometry Data Centre produce the Mass Spectrometry Bulletin which contains a worldwide citation index of

[22] K. Kenmotsu, M. Imanaka, K. Matsunaga, J. Oda, Y. Okamoto and T. Ishida, *Okayama-ken Kanko Hoken Senta Nenpo*, 1987, **11**, 89.
[23] A. A. Stevens, L. A. Moore, C. J. Slocum, B. L. Smith, D. R. Seeger, and J. C. Ireland, *Adv. Chem. Ser.*, 1989, **219**, 618.
[24] D. W. Hewetson and C. J. Mirocha, *J. Assoc. Off. Anal. Chem.*, 1987, **70**(4), 647.
[25] J. Nakamura, S. Takahashi and T. Takeda, *Shimizu, Kiyoshi, Shimadzu Hyoron*, 1986, **43**(4), 365.
[26] D. Garozzo and G. Montaudo, *J. Anal. Appl. Pyrolysis*, 1985, **9**(1), 1.
[27] S. Foti, D. Garozzo and G. Montaudo, *Chim. Ind. (Milan)*, 1983, **65**(10), 641.
[28] N. L. Bakowski, E. C. Bender and T. O. Munson, *J. Anal. Appl. Pyrolysis*, 1985, **8**, 483.
[29] M. Weisser, *Schriftenr. ISWW Karlsruhe*, 1993, **63**, 211.
[30] V. I. Vil'chinskaya, V. P. Suboch and V. D. Matveentsev, *Zh. Prikl. Spektrosk.*, 1984, **41**(6), 991.
[31] I. Horman, H. Traitler and J. Aeschlimann, *J. High Resolut. Chromatogr.*, 1989, **12**(5), 308.
[32] G. Olerich and M. V. Buchanan, report, ORNL/TM-8855; order no. DE83017887, 191pp Avail NTIS from: Energy Res. Abstr. 1983, **8**(21), Abstr. no. 51787.
[33] K. Varmuzza, 'Recent Dev. Mass Spectrom., Biochem. Med.', (Proc. Int. Symp., ed. A. Frigerio, Plenum New York, N.Y., USA, 4th, meeting date 1977, vol. 1, 371–8.
[34] J. B. Reeves III and G. C. Galletti, *Org. Mass Spectrom.*, 1993, **28**(5), 647.
[35] A. Yasuhara, H. Ito, K. Kenmotsu and J. Oda, *Kankyo Kagaku*, 1993, **3**(1), 101.
[36] D. E. Roques, E. B. Overton and C. B. Henry, *J. Environ. Qual.*, 1994, **23**(4), 851.
[37] W. P. Jacobs, *J. Phycol.*, 1993, **29**(5), 595.
[38] L. M. Libbey, *J. Essent. Oil Res.*, 1991, **3**(3), 193.
[39] J. Nowicki, *J. Forensic Sci.*, 1993, **38**(6), 1354.
[40] D. P. Moon and T. E. Madden, *Inst. Phys. Conf. Ser.*, 1987, **90**(Electron Microsc. Anal., 1987), 69.

Mass Spectrometry and related subjects, which is also available online.[41]

228 High-energy collision-induced dissociation tandem mass spectra of peptides have been produced for the low mass ions (below mass 300).[42] Another MS–MS database of 300 low energy spectra are to be found in the PC package PDNL.[43]

As all the major databases in mass spectrometry consist of electron ionization (El) mass spectra a method was published for chemical ionization (CI) mass spectra that manipulated the CI spectra enough to allow reasonable results from searching EI–MS databases.[44]

The availability of large databases for the mass spectroscopist should not be allowed to lead to a false sense of security. Many publications exist in the literature detailing just where reliance on the 'large' database fails to deliver the desired results. Quite often, task specific, smaller, sample relevant databases are of more use as in the study of the methylation of polysaccharides,[45] peptides[46] or carbohydrates.[47] Spectroscopists should be wary of the 'my database is bigger than your database' arguments, as bigger is seldom equivalent to better!

Simple database searches can also be supplemented by intelligent use of the structure information presented when identical spectra to the unknown are not avaialble in the database.[48]

A specialist database of inductively coupled plasma (ICP) and glow discharge mass spectroscopy interference data called MS Interview was developed for Apple Macintosh personal computers and has been ported to run on an IBM-PC under Microsoft Windows.[49] Another specialist database of Laser Microprobe Mass Analysis (LAMMA) spectra has been shown to aid evaluation of the results obtained using this method although poor reproducibility limits the automation.[50]

5.2 Infrared and Raman Spectral Databases

The major infrared (IR) spectral database currently available is the Sadtler IR Digital Spectra Libraries.[51] 69 000 condensed phase spectra make up the main digital library, which is expanding by *ca.* 2 000 spectra per year, and many other smaller specialist libraries are available. The database is available for installation on many spectrometer control computers or PCs running third party software such as Galactic's SpectraCalc, LabCalc or GRAMS packages or the SPECTACLE

[41] The Royal Society of Chemistry (RSC), Thomas Graham House, Science Park, Milton Rd, Cambridge, CB4 4WF, UK.

[42] A. M. Falick, W. M. Hines, K. F. Medzihradsky, M. A. Baldwin and B. W. Gibson, *J. Am. Soc. Mass Spectrom.*, 1993, **4**(11), 882.

[43] E. W. H. Hayek, W. Stanek and P. Krenmayr, *Rapid Comm. Mass Spectrom.*, 1993, **7**(1), 99.

[44] D. F. Hunt and P. J. Gale, *Anal. Chem.*, 1984, **56**(7), 1111.

[45] M. Durindova, P. Jurasek, I. Klbik and V. Kavocik, *Chem. Pap.*, 1993, **47**(3), 198.

[46] M. R. Gold, T. Yungwirth, C. L. Sutherland, R. J. Ingham, V. Daisy, R. Chiu, I. van Oostveen, H. D. Morrison and R. Aebersold, *Electrophoresis*, 1994, **15**(3–4), 441.

[47] M. Durindova, I. Klbik and V. Kovacik, *Trends Anal. Chem.*, 1992, **11**(10), 355.

[48] W. Werther, H. Lohninger, F. Stancl and K. Varmuzza, *Chemom. Intell. Lab. Syst.*, 1994, **22**(1), 63.

[49] L. L. Burton and G. Horlick, *Spectrochim. Acta, Part B*, 1993, **48**(8), E1063.

[50] L. Wouters, D. Michaud and R. Van Grieken, *Mikrochim. Acta*, 1993, **110**(1–3), 31.

[51] Sadtler IR Digital Spectra Libraries', Sadtler Division, Bio-Rad Laboratories Inc., 3316 Spring Garden Street, Philadelphia, PA 19104, USA.

software from LabControl GmbH.[52,53] One major drawback of the Sadtler data was the lack of structural information or connection tables linked to the spectral information; however this problem has recently been addressed.

The Canadian Scientific Numerical Database Service (CAN/SND) has an IR database system called Search Program for IR Spectra (SPIR). This database consists of 140 000 commercial and 26 000 literature spectra and has a 'strong' band/'weaker' band/ no peaks region template search routine.[54]

The IR section of the CIS was called the IR Search System (IRSS) and had about 6500 complete spectra obtained from USEPA contracts and the Boris Kidric Institute collection in the former Yugoslavia. Searches are possible on this system by peak position, partial or complete molecular formulae, chemical structure and CAS number. The spectrometer manufacturers offer various databases to their customers. Nicolet–Aldrich have started to include structure information in their databases. Bruker–Merck have produced the FTIR atlas. Based on a search library of 4000 spectra with a resolution of 2 cm^{-1} the atlas includes chemical structure, molecular formulae (search criteria), molecular masses, b.p./m.p., preparation information, CAS numbers and Merck numbers. They also offer a polymer library of 900 spectra, the EPA gas phase library of 2300 spectra, Georgia State Crime Labs. spectral library of 744 spectra and all the Sadtler digital libraries.[55]

The IDIOTS structure elucidation system of the BASF contained 17 000 spectra with a resolution of 2 cm^{-1} displayed in negative absorbance (transmission converted into absorbance which is then reflected around the wavenumber axis before being baseline corrected), topologically encoded structures, CAS numbers and CAS names.[56] This system is currently being upgraded and revitalized by Otto and co-workers at the Bergakademie Freiberg, Germany.

Medical diagnostics have also been carried out with the aid of dedicated IR spectral libraries. An example is the analysis of urinary calculi by IR[57] or Raman laser fibre optic spectroscopy;[58] however, an 'experienced analyst' is advised as being necessary to interpret the search results from such complicated biological systems!

Another German group reported a database of *ca.* 12 000 chemicals[59] and Zupan reported a library of polymer IR spectra.[60] A molecular line parameter database was constructed from the measurements taken during the SPACELAB 3 mission where the ATMOS instrument collected 0.01 cm^{-1} resolution solar spectra. In total 500 spectra in the range 2.3–16 μm were measured through the atmosphere.[61]

52 Galactic Industries Corporation, 395 Main Street, Salem, NH 03079, USA.
53 LabControl GmbH, Max-Planck-Str., 17a, 50858 Köln, Germany.
54 G. H. Wood, J. R. Rodgers and S. R. Gough, *J. Chem. Inf. Comput. Sci.*, 1989, **29**(2), 118.
55 Bruker Analytische Messtechnik GmbH, Karlsruhe, Germany.
56 M. Passlack and W. Bremser, 'IDIOTS – Structure oriented data bank system for the identification and interpretation of infrared spectra', in 'Computer Supported Spectroscopic Databases', ed. J. Zupan, Ellis Horwood, Pub., Chichester, UK, 1986, ch. 5.
57 G. Rebentisch, M. Doll and J. Muche, *Laboratoriumsmedizin*, 1992, **16**(6), 224.
58 T. D. N. Hong, D. Phat, P. Plaza, M. Daudon and N. Q. Dao, *Clin. Chem.*, 1992, **38**(2), 292.
59 M. Wolf, *Krim Forensische Wiss.*, 1987, **67–68**, 174.
60 M. Penca, J. Zupan and D. Hadzi, *Kem. Ind.*, 1978, **27**(3), 137.
61 L. R. Brown, C. B. Farmer, C. P. Rinsland and R. A. Toth, *Applied Optics*, 1987, **26**(23), 5154.

A library of about 500 GCIR spectra of compounds relevant to the identification of flavour components in complex mixtures has been put together paying special attention to the absence of contaminants.[62]

An IR microscopy data collection has been published in electronic form with 53 polymer spectra being published on PC disks in the Joint Committee on Atomic and Molecular Physical Data–Data Exchange (JCAMP–DX) standard data format.[63–64]

Raman spectra remotely collected can aid speed of decision making and worker safety. A fibre optic system has been demonstrated collecting FT near IR (NIR) Raman spectra of transuranic low level drum waste. Low concentrations of toxic materials in binary salt mixtures were identified but for extended sample ranges a larger solid waste library would be required.[65] The use of FT Raman for the direct analysis of TLC plates has been published along with a discussion of the best adaption of library search algorithms for this application when a medium sized Raman library is to be searched.[66]

The authors[66] describe a technique to obtain spectra directly from TLC plates and compare four different search algorithms of the difference type outlined above. All four algorithms correctly identified the five compounds tested, even from noisy spectra.

FT Raman is an area which is likely to expand because of the introduction of the near-IR laser. It therefore has the benefit of gathering data in a modern computer environment, without the restrictions of memory for software, slow CPUs and lack of storage space. We hope that this will lead to better databases, but hardware marches on; the next revolution, the fast CD-ROM, should allow millions of entries in a library and, hopefully, very fast access. The technique itself is in many ways superior to traditional IR, since sampling technique problems largely disappear (see above for mid-IR) and earlier sensitivity and fluorescence problems are now less important.

A study has been carried out in which the usefulness of searching conventional condensed phase spectral libraries with data gathered from cryotrapping GC environmental samples has been investigated.[67] The technique here is to trap the eluant from a GC column on a zinc selenide window at 77 K. The authors used conventional solid phase IR libraries and examined their efficiency for environmental samples. They commented that the results were useful, but is this technique superior to GCMS?

The ease and safety of using FTIR methods for the identification of clinical isolates was demonstrated using a database of 139 bacterial spectra.[68]

Inorganic IR collections are rare however a mineral spectra digital collection has been reported.[69]

[62] G. Fischboeck, R. Kellner and W. Pfannhauser, *Agric., Food Chem. Consum., Proc. 5th Eur. Conf. Food Chem.*, INRA, Paris, France, 1989, vol. 2, 631.
[63] M. W. Tungol, E. G. Bartick and A. Montaser, *Spectrochim. Acta*, 1991, **46(B)**, 1535.
[64] M. W. Tungol, E. G. Bartick and A. Akbar, *Appl. Spectrosc.*, 1990, **44**(4), 543.
[65] B. A. Crawford, J. R. Jewett, K. E. Parker, C. A. Petersen and C. J. Petty, *Proc. SPIE-Int. Soc. Opt. Eng.*, 1993, **2069**, 65.
[66] C. Petty and N. Cahoon, *Spectrochim. Acta, Part A*, 1993, **49A**(5–6), 645.
[67] T. Visser and M. J. Vredenbregt, *Vib. Spectrosc.*, 1990, **1**(2), 205.
[68] D. Helm, H. Labischinski, G. Schallehn and D. Naumann, *J. Gen. Microbiol.*, 1991, **137**(1), 69.
[69] J. W. Salisbury, L. S. Walter and N. Vergo, *Am. Mineral.*, 1989, **74**(7–8), 938.

One of the most often used databases is the HITRAN database of spectroscopic parameters.[70] This database can be used to calculate spectra and is often used as a reference source when reference spectra are unavailable.

5.3 Nuclear Magnetic Resonance (NMR) Spectral Databases

The CSEARCH program running on work-stations under UNIX developed by Robien has been commercialized by Bio-Rad and now contains some 60 000 assigned ^{13}C NMR spectra which are also linked to the Sadtler IR collection[71] and some 90 000 in the version of the University of Wien.[72]

In 1986 a report was published on the Japanese SPIRES database. Designed to support the CHEMICS expert system[73] it combined the NIH–EPA and Bremser Microfiche data for ^{13}C NMR (*ca.* 15 000 spectra) with ^{1}H NMR data measured by the scientists themselves (10 000 total with 8 000 in computer readable form but only 400 inspected and added to the database at the time of publication). There was a prospective overlap between the two databases of only *ca.* 1000 substances.[74] By 1989, a 2D-INADEQUATE input into the database had been made available.[75] Other reported Japanese databases are a ^{1}H NMR and ^{13}C NMR parameter database of laboratory and literature spectra.[76–78]

In 1987, Zupan described the expert system CARBON for the identification of organic compounds from their ^{13}C NMR spectra which would also simulate ^{13}C NMR spectra for new compounds. This system was built around a database of 2 536 ^{13}C NMR spectra of organic compounds.[79]

A Chinese group has reported an 'artificial intelligence' system for the interpretation of proton NMR spectra of polymers which includes a spectral database[80] as well as a ^{13}C NMR collection of different flavonoid natural products[81] and a small ^{13}C NMR database of steroidal natural products.[82] Another reported on a search system CIAC in 1986 capable of handling *ca.* 12 000 ^{13}C NMR spectra of organic compounds.[83] Also in 1986 an American ^{13}C NMR database was reported combined with a chemical reaction database by reagents, using the Molecular Access System (MACCS) and Data Acquisition System (DATACCS), used as a pre-clinical database for drug synthesis.[84] Many groups

[70] For example C P. Rinsland, A. Goldman and J. M. Flaud, *J. Quant. Spectrosc. Radiat. Transfer*, 1992, **48**(5–6), 693, and L. S. Rothman *et al. J. Quant. Spectrosc. Radiat. Transfer*, 1992, **48**(5–6), 469.
[71] B. Woods, *Am. Lab.*, 1993, **25**(19), 28.
[72] C. Lingran and W. Robien, *J. Chem. Inf. Comput. Sci.*, 1994, **34**(4), 934.
[73] K. Funatso, *Yuki Gosei Kagaku Kyokaishi*, 1993, **51**(6), 516.
[74] S. Sasaki, T. Yamasaki, K. Tanaka and H. Abe, in 'Computer Supported Spectroscopic Databases', ed. J. Zupan, Ellis Horwood Pub., Chichester, UK, 1986, p. 133.
[75] K. R. G. Funatsu, Y. Susuta and S. Sasaki, *J. Chem. Inf. Comput. Sci.*, 1989, **29**(1), 6.
[76] O. Yamamoto, K. Hayamizu and M. Yanagisawa, *Anal. Sci.*, 1989, **5**(2), 141.
[77] O. Yamamoto, K. Hayamizu and M. Yanagisawa, *Anal. Sci.*, 1988, **4**(5), 445.
[78] O. Yamamoto, K. Hayamizu and M. Yanagisawa, *Anal. Sci.*, 1988, **4**(5), 461.
[79] J. Zupan, M. Novic, S. Bohanec, M. Razinger, L. Lah, M. Tusar and I. Kosir, *Vestn. Slov. Kem. Drus.*, 1987, **34**(1), 83.
[80] L. Gao, T. Zhao, G. Wei and X. Liang, *Sci. China, Ser. B*, 1993, **36**(8), 897.
[81] L. Gao, T. Zhao and X. Liang, *Bopuxue Zazhi*, 1992, **9**(4), 307.
[82] T. Zhao, L. Gao, J. Zhang, Q. Guo and X. Liang, *Fenxi Huaxue*, 1993, **21**(6), 657.
[83] L. Xu, G. Li, S. Wang and J. Sun, *Fenxi Huaxue*, 1986, **14**(6), 431.
[84] S. Barcza, H. W. Mah, M. H. Myers and S. S. Wahrman, *J. Chem. Inf. Comput. Sci.*, 1986, **26**(4), 198.

are involved in extracting the chemical literature for ^{13}C NMR spectral information, which is then added to their own databases.[85]

A PC based ^{13}C NMR program suite called CSPEC2 includes not only standard library features but also prediction algorithms.[86]

A substantial collection of boron NMR data has been put together under the Chembase software sytstem.[87] The literature extraction is ongoing and in 1990 the reported size of the database was 9 400 units.[88]

Where gaps in the databases occur some workers try to fill them by calculating the spectra that are missing![89] Specialist databases are to be found in NMR as in the other types of spectroscopy. An example is a program developed to assist in the determination of the primary structures of carbohydrates which contains structure and ^{1}H NMR spectra[90] or polysaccharides containing ^{1}H and ^{13}C NMR data.[91] Another is a collection of protein NMR data.[92]

5.4 UV–VIS Databases

UV–VIS databases are rather rare. In 1988, a UV spectral database of polyaromatic hydrocarbons (PAHs) with physicochemical data was reported for a supercritical fluid chromatography system,[93] and a spectral library search system for identifying metabolites in urine by HPLC–UV was published.[94] This was the latest in a line of publications from scientists using HPLC–UV systems for drug analysis of one sort or another. In 1987, an HPLC–UV database system for routine drug analysis containing 300 reference spectra measured on a diode array spectrophotometer was reported,[95] as was a relatively large library of drug spectra used for the analysis of drugs in biological fluids, impurities in heroin, PAHs in oil and indole alkaloids in plant tissue measured by HPLC–UV spectrometry and diode array detection.[96]

A similar setup was reported to aid the identification of organic compounds whereby 186 sets of diode array UV spectra LC retention data from 74 organophosphorous pesticides were tested against 186 spectra of the same compounds measured under different experimental conditions,[97] and for toxin identification.[98]

85 For example – BASF; Professor W. Robien, University of Wien; Dr Haupt, University of Hamburg.
86 H. N. Cheng and L. J. Kasehagen, *Anal. Chim. Acta*, 1994, **285**(1–2), 223.
87 H. Noeth and E. Striedl, Software-Entwickl. Chem. 3, Proc. Workshops 'Comput. Chem.', ed. G. Gauglitz, Springer, Berlin, Germany, 3rd edn., 1989, meeting date 1988, pp. 101–8.
88 H. Noeth and E. S. Striedl, *Chem. Struct. 2. Proc. Int. Conf.*, ed. W. Warr, Springer, Berlin, Germany, 1993, meeting date 1990, pp. 15–32.
89 M. L. Ranc and P. C. Jurs, *Anal. Chim. Acta*, 1993, **280**(1), 145.
90 J. A. Van Kuik, K. Hard and J. F. G. Johannes, *Carbohydr. Res.*, 1992, **235**, 53.
91 P. E. Jansson, L. Kenne and G. Widmalm, *Carbohydr. Res.*, 1989, **188**, 169.
92 J. L. Markley, *et al.*, *NATO ASI Ser., Ser. A*, 1991, **225**, 39.
93 K. Jinno, M. Kuwajima and K. Masuda, *Lab. Microcomput.*, 1988, **7**(2), 52.
94 L. Hubert and K. Zech, *J. Pharm. Biomed. Anal.*, 1988, **6**(6–8), 1039.
95 D. W. Hill, T. R. Kelley and K. J. Langner, *Anal. Chem.*, 1987, **59**(2), 350.
96 D. M. Demorst, J. C. Fetzer, I. S. Lurie, S. M. Carr and K. B. Chatson, *LC-GC*, 1987, **5**(2), 128.
97 H. J. Boessenkool, P. Cleij, C. E. Goewie, H. H. Van den Broek and H. A. Van't Klooster, *Mikrochim. Acta*, 1986, **2**(1–6), 75.
98 M. I. Selala, A. Musuku and P. J. C. Schepens, *Anal. Chim. Acta*, 1991, **244**(1), 1.

A team from Sofia published information on a database of standardized optical reflectivity from minerals using the programs POLYFIT and POLYORT.[99] A triplet–triplet absorption spectral database has been compiled for 1143 compounds from the literature and is available through the University of Notre Dame, Radiation Data Centre.[100]

A small collection of UV data exists at the Institut für Physikalische und Theoretische Chemie, Tübingen, Germany.[101]

A database DATALOG from the Syracuse Research Corporation contains 12 000 records from over 8 000 organic compounds and metals. The database contains information relevant to the environmental fate of these substances and where relevant this includes the UV spectra.[102]

5.5 Other Spectral Databases

Several small databases exist for the less widely used forms of spectroscopy. A Mössbauer Spectroscopy database exists at the Mössbauer Effect Data Centre, University of North Carolina.[103] Two PIXE databases have been reported, one of 9 000 spectra from proton beam bombardment of 14 thin film target lanthanides in the energy range 1–2.5 MeV,[104] and the other, somewhat more comprehensive, of PIXE thick target spectra from proton and helium (HIXE) experiments in the range 100 keV–10 MeV for all target elements from H to U for both K and L shell *X*-rays.[105]

Nuclear quadrupole resonance (NQR) spectra have been extracted from the literature by a Japanese group. The compilation frequency data were obtained by reviewing original papers, reprints and published data, retrieving the nuclear mass number, the temperature of the measurements, NQ coupling constants, and, where available, the CAS numbers and molecular formulae.[106]

A review has been published which covered efforts to produce surface analysis databases such as *X*-ray photoelectron spectroscopy (XPS), secondary ion mass spectra and Auger data.[107]

A Circular Dichroism (CD) database is constantly being updated.[108]

The use of neural networks to handle ion mobility spectrometry (IMS) data sets has been proposed as being more relevant than conventional databases due to the lack of specific structural information in the data. A collection of some 292 IMS spectra were used for this work.[109]

[99] V. Kovachev, S. Strahimirov and D. Damyanov, *God. Vissh. Minno-Geol. Inst. Sofiya*, 1987, **33**(2), 41.
[100] University of Notre Dame, Radiation Chemistry Data Centre, Radiation Laboratory, Notre Dama, IN 46556, USA.
[101] Personal communication, Professor Gauglitz.
[102] Syracuse Research Corp., Merril Lane, Syracuse, NY 13210, USA, Technical Database Services Inc., 10 Columbus Circle, Suite 2300, New York, NY 10019, USA.
[103] J. G. Stevens, *Hyperfine Interact.*, 1988, **42**(1–4), 1215.
[104] T. Hirokawa, F. Nishiyama and Y. Kiso, *Nucl. Instrum. Methods Phys. Res., Sect. B*, 1988, **B31**(4), 525.
[105] D. D. Cohen and E. Clayton, *Nucl. Instrum. Methods Phys. Res., Sect. B*, 1987, **B22**(1–3), 59.
[106] H. Chihara, *J. Mol. Struct.*, 1982, **83**, 1.
[107] C. J. Powell, *Surf. Interface Anal.*, 1991, **17**(6), 308.
[108] N. Ikemoto, L. C. Lo, O. K. Kim, N. Berova and K. Nakanishi, *Carbohydr. Res.*, 1993, **239**, 11.
[109] S. E. Bell, W. C. Mead, R. D. Jones, G. A. Eiceman and R. G. Ewing, *J. Chem. Inf. Comput. Sci.*, 1993, **33**(4), 609.

5.6 Stellar Databases

Several databases have been produced as a result of space exploration. One has already been mentioned as it does not contain data from stellar objects. In 1987, a far-UV (123–193 nm) library of mean stellar energy distributions was published,[110] which was later updated with a mid-UV (190–320 nm) library of low resolution spectra from the International UV Explorer Atlas and new cool (EGK) star spectra.[111] A spectral solar radiation database from the Solar Energy Research Institute has also been reported.[112]

The IR astronomical satellite (IRAS) has produced a very large database covering 8–24 μm of various subjects from the low resolution spectrograph.[113] A library of 56 stellar spectra from 1.428–2.5 μm from a FT spectrograph has also been reported.[114]

6 MULTI-DIMENSIONAL SPECTROSCOPIC DATABASES

With the advances in storage space, CPU speed on PCs and cheap memory, the future of spectroscopic databases must lie in the use of combined systems where a problem can be addressed from several different angles at once with the different spectroscopic databases linked by a more or less 'expert' system. This system will be different for each technique. The effectiveness of combining GCIR and GCMS data in a single system, for example, has already been clearly demonstrated.[115] Several significant steps have already been taken in this direction in Germany and Japan.

The German Cancer Research Centre described a combined system called SPEKTREN II which includes ^{13}C NMR, mass and infrared spectra.[116] The Japanese system is somewhat larger and contains IR, NMR, mass, ESR and Raman databases using complete spectral charts, names, molecular formulae and CAS numbers.[117–118] A forerunner of the BASF system[119] was described by Bremser[120] with 42 000 structure oriented mass spectra and an interpretation system. In many cases several different spectroscopic techniques are included in general factual databases but without any direct link between the searches.

The enforcement of international regulations also drives the production of substance specific databases, as with a collection of relevant IR and mass spectra of monomers and other plastic starting material. This supports a handbook of

[110] M. N. Fanelli, R. W. O'Connel and T. X. Thuan, *Astrophys. J.*, 1987, **321**(2, Pt. 1), 768.
[111] M. N. Fanelli, R. W. O'Connel, T. X. Thuan, D. Burstein and C. C. Wu, Eur. Space Agency, (Spec Pub.) ESA SP, 1988, 281 (Decade UV Astron. IUE Satell., vol. 2), 377.
[112] C. Riordan, *Sol. Cells*, 1988, **24**(3, 4), 313.
[113] P. R. Roelfsema, T. De Jong and F. Lahuis, *Infrared Phys. Technol.*, 1994, **35**(2, 3), 439.
[114] A. Lancon and B. Rocca-Volmerange, *Astron. Astrophys., Suppl. Ser.*, 1993, **96**(3), 593.
[115] J. R. Cooper and C. L. Wilkins, *Anal. Chem.*, 1989, **61**(14), 1571.
[116] T. Förster, C. W. von der Lieth and H. J. Opferkuch, *GIT Fachz. Lab.*, 1989, **33**(4), 318.
[117] J. Hiraishi, *JETI*, 1984, **32**, 54.
[118] K. Hayamizu, *Kagaku to Sofutowea*, 1992, **14**(3), 191.
[119] W. Bremser and M. Grzonka, *Mikrochim. Acta*, 1991, **2**(1–6), 483.
[120] R. Neudert, W. Bremser and H. Wagner, *Org. Mass Spectrom.*, 1987, **22**(6), 321.

reference data for the European Community (EC) regulations on food contact materials specially aimed at helping enforcement.[121]

A group from Poland used dBase III+ running on an IBM-PC to store spectroscopic data for diatomic molecules.[122] A Chinese paper described the development of systems including spectral, chemical, engineering and chemical documentation databases[123] and although these are not directly accessible they may prove to be the basis for collaboration.

More applied combinations have been reported in the literature, a good example of which is the combination, by the Tokyo Customs Laboratories, of IR spectra and forensic Marquis test results for around 16 000 drugs and narcotics in an analytical reference system.[124] The usefulness of combined GC–IR–MS systems in forensic science has been tested in the analysis of human urine for drug abuse determinations involving amphetamine type compounds.[125] L'Oreal Cosmetics have reported an IR ^{1}H NMR and ^{13}C NMR spectra system of polymers used in cosmetics.[126] Dupont have also developed a PC based multi-spectral database system containing ^{1}H, ^{13}C, ^{15}N, ^{19}F, ^{29}S and ^{31}P NMR as well as IR and UV–VIS databases with structural information relevant to materials encountered in the coatings and plastic industry.[127]

A ^{13}C NMR and IR database was available through STN online. The collection was a combination of BASF, Gesellschaft für Biotechnologische Forschung (GBF) and the Imperial Chemicals Industries plc (ICI) spectra. The system contained 83 000 ^{13}C NMR spectra from 67 000 compounds and 8 500 IR spectra from 8 300 compounds. Recently the system has been upgraded and is now available with new databases and a large range of new functions including spectra prediction and cross referencing with the Beilstein, CAS and other databases on STN.[128,129] The new system called SpecInfo Online is best accessed *via* the STN Express PC software package where the most use can be made of the off-line structure editing facilities.

The SpecInfo system is probably the largest commercial multi-dimensional spectroscopic data collection in the world with currently 134 520 NMR, MS and IR spectra of 107 000 structures.[130]

A system called SCANNET has been reported which offers ^{13}C NMR, ^{1}H NMR, IR, Raman, mass spectra, and UV spectra databases.[131–133]

An EU project has recently aimed to collect FTIR, NMR and NIR data of

[121] J. Gilbert, J. Bush, A. Lopez de Sa, J. B. H. Van Lierop and X. Coenaga, *Food Addit. Contam.*, 1994, **11**(1), 71.

[122] N. Badowski and W. Zyrnicki, *Pr. Nauk. Inst. Chem. Nieorg. Metal. Pierwiasttkow Rzadkich Politech. Wroclaw*, 1987, **58**, 60.

[123] X. Yu, *Jisuanji Yu Yingyong Huaxue*, 1987, **4**(1), 24.

[124] S. Inde, Y. Uno, M. Ishiguro and N. Shigeru, *Kanzei Chuo Bunsekishoho*, 1987, **27**, 127.

[125] G. E. Platoff Jr., D. W. Hill, T. R. Koch and Y. H. Caplan, *J. Anal. Toxicol.*, 1992 **16**(6), 389.

[126] P. Gilard, C. Pope and P. Bore, *Cosmet. Sci. Technol. Ser.*, 1985, **4**, 289.

[127] K. Adamson, *J. Appl. Polym. Sci.: Appl. Polym. Symp.*, 1992, **51**, 263.

[128] A. J. Barth, *J. Chem. Inf. Comput. Sci.*, 1993, **33**, 52.

[129] D. Canzler, M. Hellenbrandt and Fresenius, *J. Anal. Chem.*, 1992, **344**(4, 5), 52.

[130] M. Penk, *Chemical Concepts*, Postfach 10 02 02, 69442 Weinheim, Germany.

[131] B. Debska, *J. Mol. Struct.*, 1993, **294**, 17; *Anal. Chim. Acta*, 1992, **265**(2), 201.

[132] B. Debska and Z. S. Hippe, *J. Mol. Struct.*, 1992, **267**, 261.

[133] B. Guzowska-Swider and Z. S. Hippe, *J. Mol. Struct.*, 1992, **275**, 225.

foodstuffs and the associated food related factual information.[134] To facilitate exchange of information between contributors to the project a common JCAMP–DX format was devised with software to convert from many manufacturers formats to the common one.

7 CONCLUSIONS

The future of spectroscopic databases is very bright. The distribution of large collections has become easier with the introduction of fast personal computers and the rapid increase in the volume of data which can be stored at very small cost.[135,136] As these collections become more expansive and more widely available the feedback from this increased use will help to weed out the incorrect data sets that plague many of the larger collections which have been built on the principle 'more is better'. The major hurdle to expanding the collections still lies in funding. Despite the obvious benefits to industry and the scientific community at large from extensive well thought out and up-to-date high quality reference spectra collections the act of coordinating and physically gathering the data sets is not glamorous enough to attract the long term stable financing that is required. Attempts are now underway to attract strategic financing for such a collective action from the European Union. What we desperately need in spectroscopy is to introduce a system similar to that run by crystallographers who must submit their crystal structure data to a central reference library which is accessible to all before any reputable journal will publish their data! This is peer review in the ultimate sense.

In conclusion, we should see more intelligent commercial systems appearing using multi-dimensional approaches to solve spectroscopic/structure problems.[137]

[134] D. N. Rutledge and P. McIntyre, *Chemom. Intell. Lab. Syst.*, 1992, **16**(1), 95.
[135] K. Tanabe, K. Hayamizu and S. Ono, *Anal. Sci.*, 1991, **7**, 711.
[136] Y. Fujiwara, M. Uetsuki and H. Chen, *Anal. Sci.*, 1991, **7**, 1601.
[137] G. C. Levy, *Spectroscopy*, 1991, **6**(4), 50. Molecular Spectroscopy: Experimental and Computing Techniques. Chapter 11 – Spectroscopic Databases.

CHAPTER 5

The Networking of Infrared Spectrometers and Computers

J. R. DIXON, W. O. GEORGE and Rh. LEWIS

1 INTRODUCTION

This chapter discusses three examples that illustrate the computer networking of IR spectrometers:

1 The use of networking to establish intensity and wavenumber standards
2 The 'searching' of a spectrum against a library of spectra
3 Molecular modelling and IR spectral correlations

Before considering these applications of networking, it is useful to summarize the ways in which digitized spectral information is currently transferred from one spectrometer (or computer) to another.

The JCAMP.DX protocol[1] permits spectra to be imported from, and exported to, spectrometers which incorporate the necessary computer hardware and software to accept files in standard format. The file transfer may be effected by:

1 Disk (sometimes known as 'diskettes')
2 An online local area network (LAN)
3 Transfer between remotes sites using JANET/BITNET links

Spectra also may be transferred to 'non-spectrometer' based packages located in the laboratory or within other PCs which may also be portable ('laptop' or 'notebook') for ease of use at locations beyond the laboratory or office for the solution of analytical, structural or other problems. Here, the screen or visual display unit (VDU) may be simultaneously projected onto a liquid crystal display (LCD) which will project onto a large screen suitable for presentation to, and discussion with, a large audience. Hence modern hardware and software developments have a potentially revolutionary impact on teaching, provided the rigour and relevance of the teaching content is not sacrificed on the media altar!

The use of the JCAMP.DX protocol permits the transfer and use of data by the

[1] R. S. McDonald and P. A. Wilks, *Appl. Spectrosc.*, 1988, **42**, 151.

problem solver. The data need not be instrument or laboratory specific and can be transformed into information with a multitude of purposes.

The issue of whether transfer is best effected by disk, LAN or electronic mail depends on the available technology and the particular user requirements. In general, transfer by supplying information on disk costs nothing over and above the costs of the PC based system. LAN and electronic mail have considerable direct and indirect costs but if in place can improve efficiency. Users are advised to retain the option of transfer of disk in the event of failure or interruption of wider systems for reasons beyond the control of the user.

2 INTENSITY AND WAVENUMBER STANDARDS

Calibrations of the intensity and wavelength of IR absorptions are complicated by the relatively narrow linewidths involved, but attempts to calibrate wavenumber values have been well established for many years. The Commission on Molecular Structure and Spectroscopy of the International Union of Pure and Applied Spectroscopy (IUPAC) provided detailed tables of IR spectra of standard compounds in 1961.[2] Since the measured wavenumber values were influenced by apparent band shape (which was in turn affected by instrumental resolution), the spectra of standard compounds in all phases were listed both for moderate or high resolution conditions using grating instruments and also for the lower resolution achieved using prism instruments. The initial tables were revised to cover extended ranges (4350–1 cm^{-1}) in 1973[3] and in 1977.[4] The values provided for wavenumber calibration have been approved by the Inter-Union Commission on Spectroscopy of the International Council of Scientific Unions (ICSU) on which International bodies for Astronomy and Physics are represented.

Recently, an analogous IUPAC project for the calibration of intensity values was coordinated by Bertie, Keefe and Jones of the University of Alberta based on earlier work at the National Research Council Laboratories, Ottawa. The outcome of this project has been published as an IUPAC Report and as a substantial monograph.[5] The work of the IUPAC Working Party chaired by Jones involved taking a number of organic liquids with suitable properties for standard intensity measurements for a number of selected bands. These comprised benzene (15 bands), toluene (8 bands), chlorobenzene (12 bands) and dichloromethane (8 bands). These liquids were chosen on the basis of their chemical and spectroscopic suitability including consideration of availability in high purity, simplicity of molecular structure, amenity to vibrational analysis and absence of conformational or hydrogen bonding equilibria which would have accentuated temperature effects because of more complex band shapes.

[2] IUPAC Commission of Molecular Structure and Spectroscopy,'Tables of Wavenumbers for the Calibration of Infra-redSpectrometers', Butterworth, London, 1961.

[3] A. R. H. Cole, R. N. Jones and R. C.Lord, *Pure and Appl. Chem.*, 1973, **33**, 605;647.

[4] A. R. H. Cole, 'Tables of Wavenumbers for the Calibration of Infra-red Spectrometers', Pergamon Press, Oxford, 2nd edn., 1977.

[5] J. E. Bertie, C. D. Keefe and R. N. Jones,'Tables of Intensities for the Calibration of Infrared Spectroscopic Measurements in the Liquid Phase', International Union of Pure and Applied Chemistry, Blackwell Scientific Publications, Oxford, 1995.

Table 1 *JCAMP.DX file of benzene taken on a dispersive Perkin-Elmer spectrometer*

```
##TITLE=BENZENE 1100–905 RUN IN 0.025 MM CELL (IRYTRUE)
##JCAMP-DX=4.24
##DATA TYPE=IR SPECTRUM
##ORIGIN=Perkin-Elmer Model 881 IR Spectrometer
##OWNER=SHAUN O'NEILL
##DATE=02-09-1990; TIME=14:26:37
##SPECTROMETER/DATA SYSTEM=PERKIN-ELMER 880
##INSTRUMENT PARAMETERS=SLIT PROGRAM=3; NOISE FILTER=3
##DATA PROCESSING=ACCUMULATIONS=1
##DELTA-X=-1
##X-UNITS=1 CM-1
##Y-UNITS=TRANSMITTANCE
##X-FACTOR=1.00
##Y-FACTOR=0.00005
##FIRST-X=1100
##LAST-X=904
##N-POINTS=197
##MIN-Y=.15405
##MAX-Y=.9004
##FIRST-Y=.89
##XY-DATA=(X++(y..y))
1100.0+17800+17806+17810+17817+17818+17819+17815+17813+17814+17814
1090.0+17805+17791+17781+17776+17769+17762+17760+17751+17737+17724
1080.0+17715+17705+17691+17678+17654+17629+17602+17580+17553+17526
1070.0+17489+17449+17411+17378+17343+17290+17227+17158+17085+17000
1060.0+16906+16804+16688+16565+16421+16258+16070+15863+15626+15347
1050.0+15021+14643+14204+13688+13080+12347+11463+10413+ 9199+ 7860
1040.0+ 6505+ 5263+ 4242+ 3523+ 3142+ 3081+ 3306+ 3794+ 4511+ 5405
1030.0+ 6417+ 7484+ 8544+ 9554+10476+11294+12008+12628+13154+13596
1020.0+13968+14277+14533+14735+14897+15017+15096+15128+15124+15115
1010.0+15144+15223+15425+15685+15921+16060+16167+16225+16272+16312
1000.0+16360+16378+16391+16370+16358+16346+16340+16339+16359+16404
 990.0+16457+16521+16575+16625+16656+16685+16714+16744+16770+16779
 980.0+16792+16797+16806+16801+16806+16800+16799+16803+16818+16822
 970.0+16834+16854+16885+16922+16961+17003+17039+17084+17126+17189
 960.0+17232+17279+17316+17369+17393+17431+17466+17510+17543+17575
 950.0+17596+17616+17637+17669+17696+17708+17726+17762+17781+17786
 940.0+17797+17819+17833+17852+17866+17888+17899+17906+17912+17927
 930.0+17931+17937+17943+17948+17950+17964+17978+17983+17973+17981
 920.0+17998+18007+17999+17991+17996+18002+18008+17999+17995+17999
 910.0+17998+17990+17990+18002+17997+17992+17991
##END=
```

As part of the project a number of workers have recorded the IR spectra of benzene in various laboratories and on a variety of spectrometers. In our laboratory we have recorded the spectrum of benzene between 1100–910 cm^{-1} on Perkin-Elmer, Mattson and Digilab spectrometers as pure liquid at a path-length of 0.025 mm. Here it may be noted that the Digilab and Mattson spectrometers are FT whilst the Perkin-Elmer is dispersive. The project also requires the spectra of interference fringes to be recorded.

Each of these spectrometers incorporates a different form of the JCAMP.DX file transfer protocol and the three versions are shown in Tables 1, 2 and 3.

Comparison of these tables shows that the layout is considerably different in each case. Nevertheless, the important information to the practitioner is the value

Table 2 *JCAMP.DX file of benzene taken on a Mattson FT spectrometer*

```
##TITLE=benzene 0.025 mm cell
##JCAMP_DX=4.24
##OWNER=
##ORIGIN=
##DATA TYPE=IR SPECTRUM
##Y-UNITS=ABSORBANCE
##X-UNITS=1 cm-1
##FIRST-X=900
##FIRST-Y=.00333801
##LAST-X=1105
##X-FACTOR=1
##Y-FACTOR=.0001
##DELTA-X=1.92847
##N-POINTS=107
##XY-DATA=(X++(Y..Y))
900.000     33    35    40    38    30    29    26    36    52    41    24
921.274     22    23    39    50    38    43    50    57    80    90    86
942.547     88    97   115   143   152   166   185   189   230   266   272
963.821    300   324   344   373   379   376   372   363   372   396   419
985.094    435   445   467   508   528   527   524   514   529   574   638
1006.368   727   821   880   900   926   997  1099  1237  1475  1850  2395
1027.642  3169  4258  5643  7036  7782  7099  5385  3770  2652  1933  1466
1048.915  1156   931   762   653   571   504   442   379   344   334   310
1070.189   278   256   232   228   235   218   199   191   190   189   191
1091.462   195   179   152   162   191   184   181   178
##END=
```

of absorbance at every wavenumber in the interval measured and it is highly desirable that different instruments closely approach some absolute value for these qualities after taking into account any scaling factors.

To determine how closely the three instruments approach a standard value the three JCAMP.DX files were read into one of the standard PC based packages which accepts the JCAMP.DX format. In our case this was the Sprouse Scientific Systems Inc. package. This converts the file from a DX file, which is an ASCII text file, into a DT file, which is a binary file. The disadvantage of the DT file is that it is not readable as text but the advantage is that it only takes about a third of the disc space.

In addition to the JCAMP, much of the work of the IUPAC project carried out in other laboratories used Galactic Industries routines to read and write different formats with the SPECTRACALC and GRAMS programs converting spectral data from the instrument format into Spectra calc. SPC format.

The graphical output to the numerical input is shown in Figure 1 which includes separate spectra of the same band for benzene in the same cell for three different spectrometers. It was established by comparison within the IUPAC working party that the measured values obtained using the FT instruments agree very closely with the average values from a number of other groups of workers, with data from the (uncalibrated) dispersive instrument differing seriously from the remaining measurements. It may therefore be concluded that for these purposes the fact that

Table 3 *JCAMP.DX file of benzene taken on a Digilab FT spectrometer*

```
##TITLE=BENZENE 0.025 mm CELL. BAND RANGE 1100–905 cm01
##JCAMP-DX=4.10
##DATA TYPE=IR SPECTRUM
##ORIGIN=BIO-RAD DIGILAB FTS50
##OWNER=SHAUN O'NEILL
##DATE=90/02/09
##TIME=13:07:04
##CREATED=fRI fEB 09 13:07:04 1990
##SPECTROMETER/DATA SYSTEM=Digilab Data system=3250
##X-UNITS=1 cm-1
##y-UNITS=ABSORBANCE
##RESOLUTION=4.0
##FIRST-X=9.02683764E+02
##LAST-X=1.10135135E+03
##FIRST-Y=-1.140267774e-02
##MAX-Y=8.430289328e-01
##MIN-Y=-2.213525400e-02
##X-FACTOR=1.92881146e+00
##Y-FACTOR=1.862645149e-09
##N-POINTS=104
##XY-DATA=(X++(Y..Y))
468-6121766-6528578-6991656-7371402-7758902-8134432-8588962-9065912-9468926
477-9851472-10115622-10388762-10747136-11044434-11284876-11528954-11702978
485-11747350-11844452-11883774-11734282-11520162-11343218-11209842-10971858
493-10582016-10130006-9549656-8789154-7924348-6942524-5799222-4563398-3508236
502-2866934-2430346-2353074-2476424-2400928-2380850-2290766-2023100-1609692
511-835870 509046 2368794 3829752 4145582 3963760 4044556 4564634 5610604
520 7537924 10778000 15701046 21241542 23826166 23273704 23416980 26081706
528 31570804 40505622 54381564 75810522 108636692 158035770 229025060
535 321125922 414939022 452597712 381859334 266070032 172599276 111686866
542 74208942 50927972 35923094 25685474 18546888 1356804 9822122 7014350
550 4872016 3172896 1778640 566638-512162-1387046-2126150-2782530-3261706
559-3606274-3965792-4284546-4604918-4843236-4987570-5145748-5308050-5446376
568-5590840-5751124-5829130--5780230
##END=
```

the two FT instrument companies concerned used different forms and versions of the JCAMP.DX protocol was not significant.

The FT results for benzene from our laboratory were incorporated into those of the other eight members of the working party and those from a number of other collaborators to produce a published set of standard tables of IR intensities.[5]

The quantities used in discussing spectral intensity merit careful definition. The basic spectroscopic measurement is the transmittance, $t(\bar{\nu})$, of the sample. This is the ratio $[I(\bar{\nu})/I_0(\bar{\nu})]$ of the monochromatic radiant power transmitted through the sample to that incident on the sample, after appropriate correction has been made for all losses other than absorption.[6] The decadic absorbance $A(\bar{\nu})$ is computed as

$$A(\bar{\nu}) = \log_{10}\left(\frac{I_0(\bar{\nu})}{I(\bar{\nu})}\right) = -\log_{10} t(\bar{\nu})$$

[6] J. E. Bertie, C. D. Keefe and R. N. Jones, *Can. J. Chem.*, 1991, **69**, 1609.

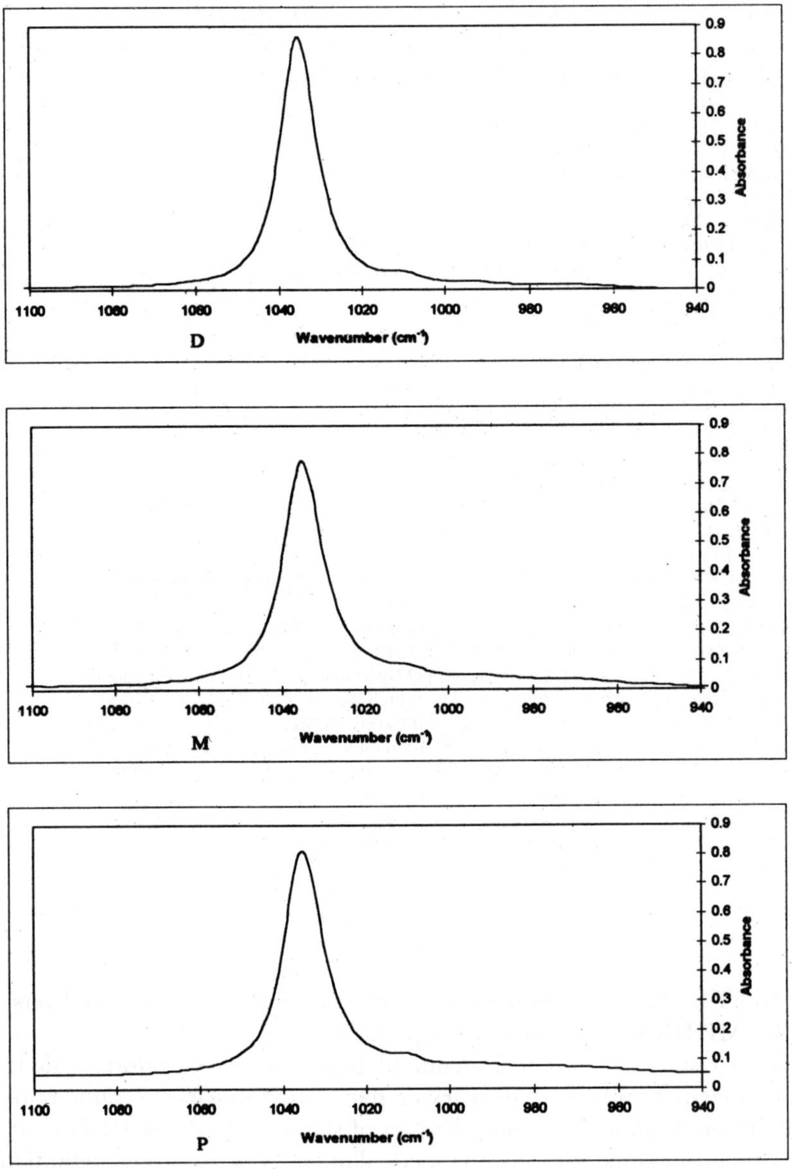

Figure 1 *Spectrum of benzene recorded on a Digilab FT Spectrometer (D), a Mattson FT Spectrometer (M) and a Perkin-Elmer Dispersive Spectrometer (P). (Cell path length* = 0.025 mm.) *NB These spectra were converted to JCamp.DX format and imported into an EXCEL spreadsheet after further conversion into two column format*

Assuming the Beer–Lambert law, several different quantities that describe the absorption intensity can be calculated from the absorbance.

Rigorous treatment of spectroscopic information requires the expression of IR absorption intensities in terms of the optical constants $n(\bar{\nu})$ and $k(\bar{\nu})$, which are respectively the real and imaginary components of the complex refractive index $\hat{n}(\bar{\nu})$ where

$$\hat{n}(\bar{\nu}) = n(\bar{\nu}) + ik(\bar{\nu})$$

The significance of these quantities can be derived directly from Maxwell's equations for the electromagnetic field.[7] Together they define precisely the optical behaviour of the sample. The imaginary component is known also as the absorption index and is given by

$$k(\bar{\nu}) = 2.303\frac{A(\bar{\nu})}{4\pi\bar{\nu}d}$$

where d is the path length and 2.303 is the conversion factor from decadic to natural logarithms. The units of d are the reciprocal of the units of $\bar{\nu}$, so that $k(\bar{\nu})$, like $A(\bar{\nu})$, is dimensionless. It is instructive to note that the linear (decadic) absorption coefficient,

$$K(\bar{\nu}) = \frac{A(\bar{\nu})}{d}$$

is a measure of the absorption by the sample in unit length, and $k(\bar{\nu})$ is, apart from numerical factors, the corresponding measure of absorption by the sample in the vacuum wavelength of the radiation.

The real and imaginary components of the complex index are related by the Kramers–Kronig integral *viz.*:-

$$n(\bar{\nu}') = n_{\infty} + \frac{2}{\pi}P\int_0^{\infty}\frac{\bar{\nu}k(\bar{\nu})}{\bar{\nu}^2 - \bar{\nu}'^2}d\bar{\nu}$$

Here it should be noted that the principal value of the integral must be taken to avoid the singularity at $\bar{\nu}'$.

It is seen that this equation carries an implication that the real refractive index be known at the infinite wavenumber. In practice, for integration over the IR region the refractive index at 8000–9000 cm^{-1} can be used, or an alternative form of the Kramers–Kronig integral can be used with a value of the refractive index at one selected wavenumber within the integration range.[8,9] It should be noted that $n(\bar{\nu})$ is dimensionless.

[7] B. Crawford Jr., T. G. Goplen and D. Swanson, in 'Advances in Infrared and Raman Spectroscopy, ed. R. J. H. Clark and R. E. Hester, Heyden, London, 1978, vol. 4, ch. 2, pp. 47–64.

[8] D. G. Cameron, D. Escolar, T. G. Goplen, A. Nadeau, R. P. Young and R. N. Jones, *Appl. Spectrosc.*, 1980, **34**, 646.

[9] T. G. Goplen, D. C. Cameron and R. N. Jones, *Appl. Spectrosc.*, 1980, **34**, 652; 657.

Two other intensity quantities, of main interest in molecular spectroscopy are ε' and ε'', the real and imaginary components of the complex dielectric constant. ε' and ε'' are also known as the dielectric permittivity and the dielectric loss, respectively. Since they are derived explicitly from n and k *viz*:

$$\varepsilon' = n^2 - k^2$$

$$\varepsilon'' = 2nk$$

they are not incorporated into the Tables of the report,[5] but provision to evaluate them from the $n(\tilde{\nu})$ and $k(\tilde{\nu})$ spectra is included in the IRYTRUE software package as the program EPSILON.

For practical purposes, the use of the rigorous units described hitherto is not commonplace and it has been customary in laboratory analytical applications for materials in the liquid phase to be characterized in terms of the molar (decadic) absorption coefficient E_m where

$$E_m(\tilde{\nu}) = \frac{A(\tilde{\nu})}{Cd} = \frac{1}{2.303}\frac{4\pi\tilde{\nu}k(\tilde{\nu})}{C}$$

in which C is the molar concentration. In the published data[5] all intensities are listed as n, k and E_m. It has been customary to use the path length d in cm, and the molar concentration in mol dm^{-3}. Accordingly, the E_m values have the unit dm^3 mol^{-1} cm^{-1}. If recommended conventions for units are followed, the path length d will be in mm, the molar concentration will be mol dm^{-3} and the E_m values will be in m^2 mol^{-1}.

Turning to the accuracy of these results, over the early period of the measurements the limiting accuracy was established as about 0.1% transmittance or 0.3% absorbance. Because of external factors, such as the effect of sample cell insertion leading to defocusing and reflection at the cell windows, the accuracy of reported intensities are somewhat less. The advent of Fourier Transform (FT) spectrometers has led to measured improvements in accuracy over dispersive grating spectrometers and the IUPAC study suggested that the accuracy of band intensities measured as band areas was within $\pm 2\%$.

To encourage uniformity in the application of these standards in quantitative analysis the working party developed an algorithm that was incorporated into a computer program named IRYTRUE and is available with the published report as a disk, which runs within standard operating systems.

To perform the calibration, IRYTRUE accepts as input a file of experimental wavenumber – absorbance values for one of the standard calibration bands. First it checks and corrects for baseline displacements across the integration range, and computes the E_m spectrum on the basis of an estimated cell path length. It then integrates the area under this E_m curve above a linear base line between the ordinate values at the integration limits. This area is ratioed to the corresponding area of the standard band, which is contained in a file accessory to the program.

IRYTRUE next computes the first approximation to a 'cell constant' by multiplying the initial cell path length by this ratio, and computes a new set of E_m

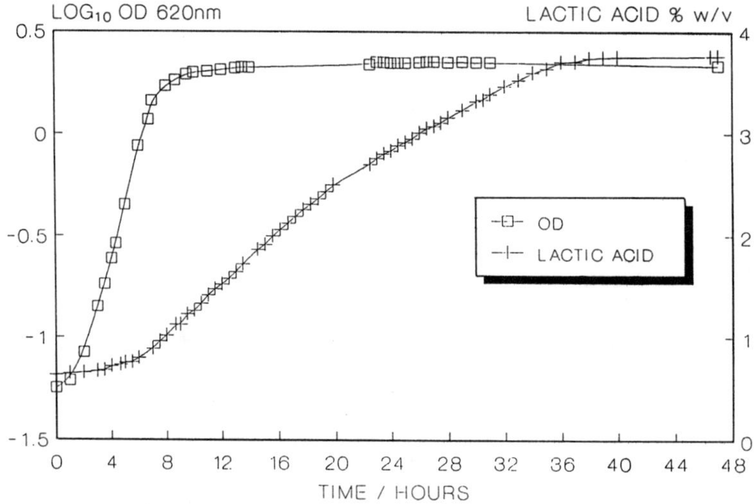

Figure 2 *Relation between growth of* Lactobacillus helveticus *and lactic acid production*

values using this cell constant in place of the path length. This procedure is iterated until a constant value for the cell constant is achieved. Usually two to three iteration cycles are sufficient. This cell constant can then be used in place of the measured cell path length in subsequent analytical calculations.

IRYTRUE also monitors the residual difference between this final E_m spectrum and the standard over the full integration range. Since the above procedure has assumed that the areas are the same, this residual error spectrum emphasizes differences in the band profile and provides a diagnostic information about the operating conditions under which the experimental spectrum was measured. If desired, this procedure can be repeated using calibration bands in different regions of the spectrum, to check the value of the cell constant.

3 APPLICATIONS OF INTENSITY MEASUREMENTS

Applications of intensity measurements require precise but not necessarily accurate measurements of intensity. Several examples will illustrate the value of such measurements.

Several computer packages were used for determination of concentration of samples taken from a fermentation process by Fairbrother *et al.*[10] The process consisted of fermenting cheese whey in an effluent from a cheese factory to lactic acid. It is possible to determine the concentration of the fermenting agent by UV spectroscopy, and of the product (lactic acid) by FTIR at fixed time intervals. A typical run is shown in Figure 2.

A similar problem concerned the online analysis of the carbon dioxide/carbonate ion formed during the anaerobic digestion of ice cream waste. In this case, FTIR

[10] P. Fairbrother, W. O. George and J. M. Williams, *Applied Microbiol.*, 1991, **35**, 301.

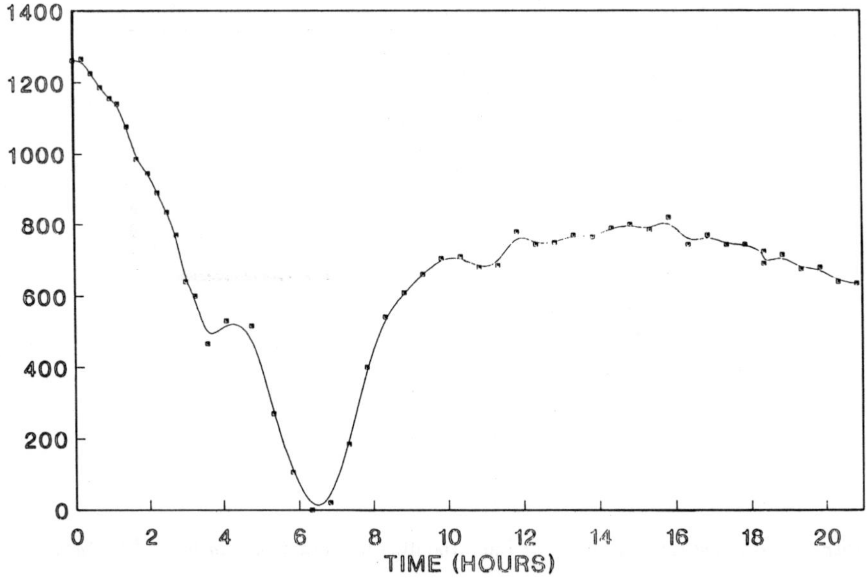

Figure 3 *Online FTIR analysis of CO_3^{2-} during the anaerobic digestion of ice cream waste at* 1360 cm^{-1}

studies[11] on aqueous samples in an ATR 'boat' cell at 1360 cm^{-1} enabled carbonate ion concentrations to be monitored with time (Figure 3). Software is under development to control the conditions of the fermenter automatically by feed-back from these analytical results.

An example outside the IR spectral region is provided by the application of atomic absorption spectroscopy taken from a study of heavy metals in the Bristol Channel by O'Boyle.[12] Here various toxic heavy metal concentrations in the Penarth area of the South Wales coast are shown as a function of distance from the shore line (Figure 4).

4 SPECTRAL SEARCH PROCEDURES

In principle, qualitative analysis (identification) may be carried out by group frequency correlations or by comparison with libraries of known compounds or by a combination of the two approaches. These procedures may be manual or computer based. In the latter case the systems may be an expert system employing computer based artificial intelligence or may use a simple search algorithm. Modern spectrometers are usually marketed with a computer system, typically a standard PC, which runs software with many powerful features including spectral manipulation facilities, quantitative analysis and spectral search facilities with computer based libraries of spectra. The JCAMP.DX (or equivalent Galactic software) feature permits spectral transfer between different instruments.

[11] A. Guwy, postgraduate project, personal communication.
[12] M. O'Boyle, undergraduate project, personal communication.

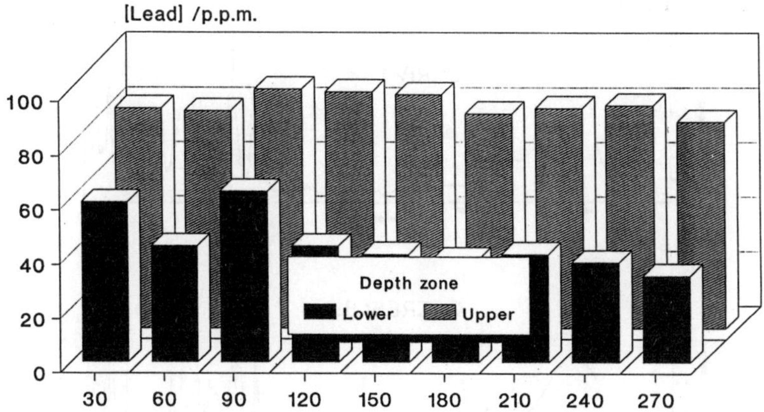

Figure 4 *Study of lead at Penarth flats*

There are also freestanding packages which are general and not instrument specific. A typical and inexpensive software package in our laboratory contains a general organic library named QLIB8AS, a polymer library, a US EPA library of 3500 spectra of gas pollutants and several 'in-house' libraries. The space occupied by these libraries is 11 Mbytes. This is stored in binary format for economy of space.

If a disk contains a directory of spectra recorded in the laboratory and converted into JCAMP.DX format, any of the spectra may be selected and searched against any library contained in the PC based package. As an example, a spectrum of a compound prepared in the laboratory with the filename DERIV.1, is shown within the Sprouse package and searched against the QLIB8AS library. The search routine selects spectra in decreasing probability of match as measured by the 'hit quality index', HIQ, and is shown in Figure 5.

It is noted that the first four predicted identities in decreasing probability are

1 *o*-cresol
2 phenol
3 2-ethylphenol
4 2-chlorophenol.

Hence the search program is confident that the sample belongs to the phenol family. Lower likelihood candidate compounds are also mainly substituted phenols (Table 4).

However, visual inspection of the spectra displayed does not support the most probable computer prediction in that the fourth compound, 2-chlorophenol, shows a better visual match than the first three. To confirm the match, any region with detailed structure may be subjected to the ZOOM command for a selected narrow spectral region and this is shown in Figure 6 together with the computer print-out of the molecular structure which is available for any of the 1200 compounds in the QLIB8AS library.

Why is the computer search less reliable than the eye of the spectroscopist? One reason lies in the variation of band contours with sample concentration or phase.

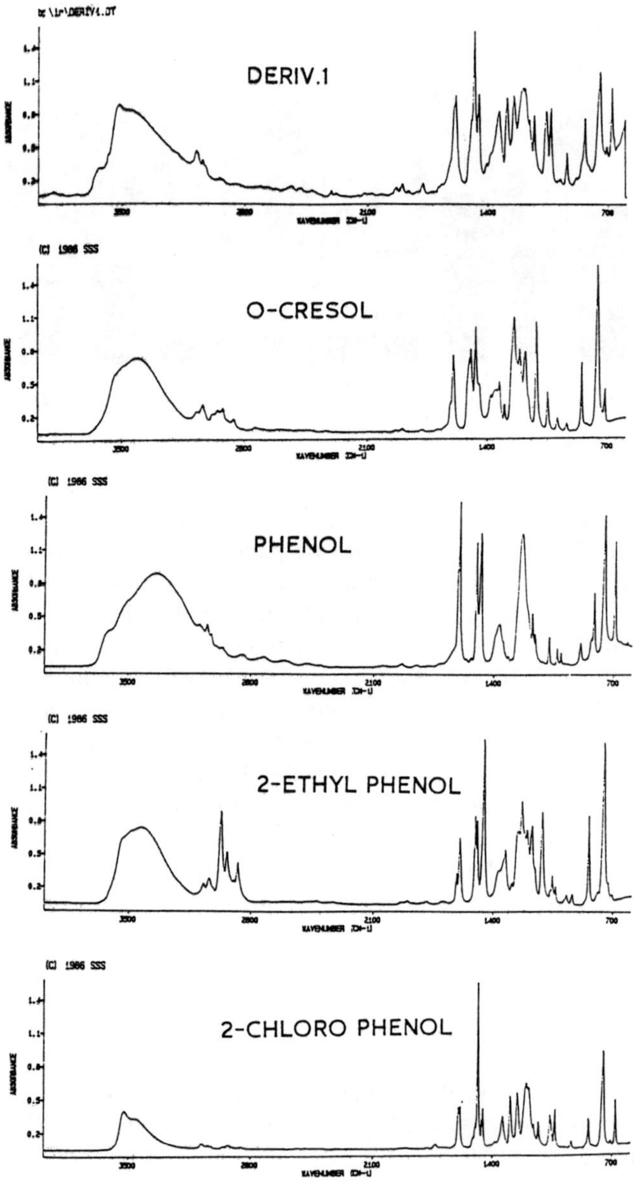

Figure 5 *Spectral search for Deriv.*1. *Search results and the corresponding spectra are presented in decreasing probability*

This leads us to expect mismatches between (for example) library spectra recorded as KBr disks and sample spectra obtained in solution. Such differences are even more accute between DRIFT and ATR reflectance spectra and libraries containing spectra recorded in transmission and separate spectra libaries are required in these cases.

Table 4 *Spectral search for unknown compound – output of 15 most likely candidates*

Search Report DB[a]: QLIB8AS *Field*: 1
Unknown: Derivl.dt *Type*: Euclid Dist
3996–700 cm^{-1}
Hits: 500

Hit	*Entry*	*HQI*	*Text*
1	276	0.20252	*o*-cresol $C_7H_8O_1$ neat
2	820	0.25332	2-ethylphenol (99%) $C_8H_{10}O_1$ neat
3	175	0.27638	2-chlorophenol $C_6H_5Cl_1O_1$ neat
4	188	0.27664	Phenol (89.3% liquified) $C_6H_6O_1$ neat
5	760	0.31015	2,6-dimethylphenol $C_8H_{10}O_1$ melt
6	431	0.31480	2,4-dimethylphenol $C_8H_{10}O_1$ melt
7	166	0.33330	*m*-cresol $C_7H_8O_1$ neat
8	297	0.34282	3-methoxyphenol $C_7H_8O_2$ neat
9	840	0.34579	2-nitrophenol (99%) $C_6H_5N_1O_3$ melt
10	834	0.35226	2-isopropylphenol (99.9+%) $C_9H_{12}O_1$ neat
11	771	0.38003	4-*tert*-butyl catechol $C_{10}H_{14}O_2$ chloroform/cast
12	827	0.38346	3-ethylphenol (99%) $C_8H_{10}O_1$neat
13	860	0.39230	3-chlorophenol (99.7%) $C_6H_5Cl_1O_1$ neat
14	828	0.40607	4-ethylphenol (99%) $C_8H_{10}O_1$ melt
15	1085	0.41046	3-isopropylphenol (tech) $C_9H_{12}O_1$ neat

[a] Data copyrighted by Sprouse Scientific Systems, Inc.

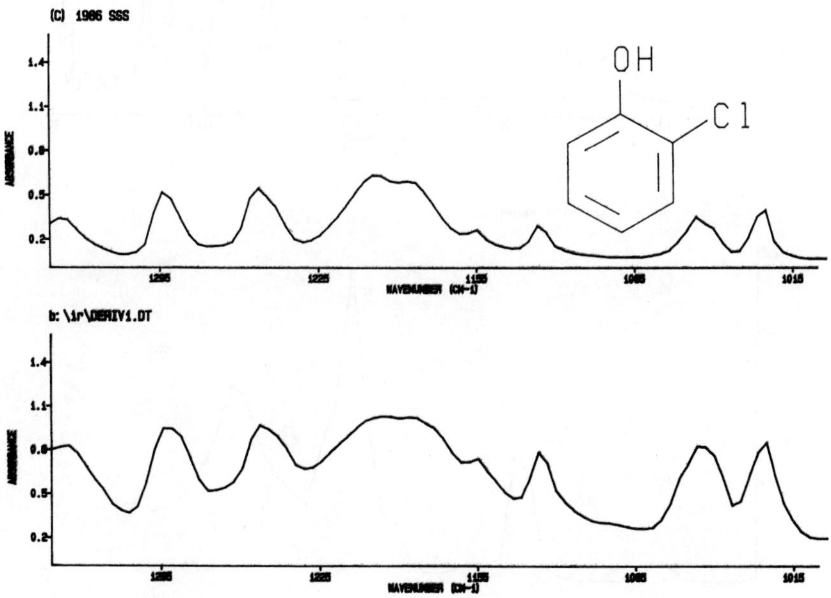

Figure 6 *The spectra of* 2-*chlorophenol (library compound) and Deriv.*1. *(the unknown) examined using the zoom facility*

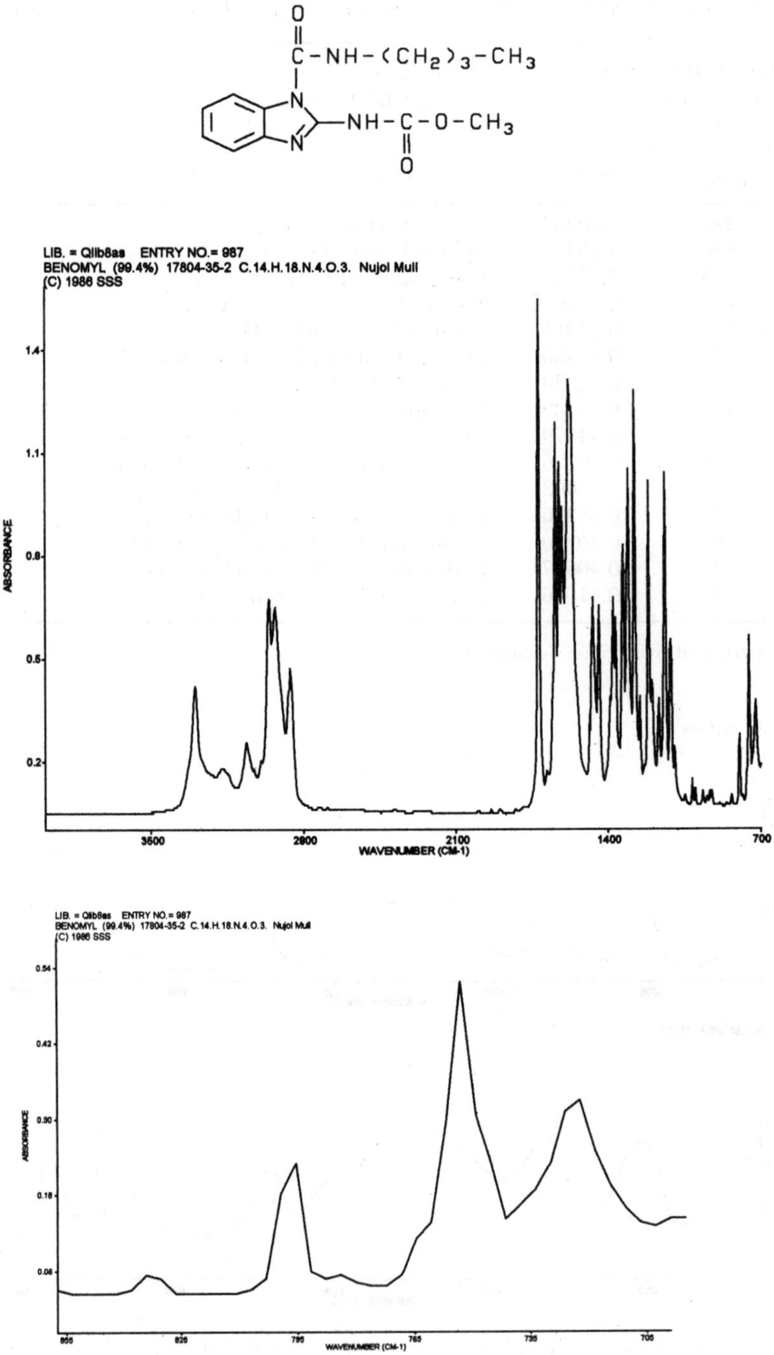

Figure 7 *The IR spectrum of benomyl,* 4000–700 cm^{-1} *and* 850–705 cm^{-1}

A second factor affecting the reliability of computer searches relates to the presence of impurities and solvent peaks which may interfere with the identification of the major component. Examples of such impurities include KBr discs which contain traces of water and weakly absorbing solutions of compounds prepared in CCl_4 solution. The trained spectroscopist will recognize such features, but it is difficult to allow for them in spectral searches because the degree of interference is variable and depends both upon sample concentration and the intensity of key bands in the spectrum of the major component.

Such interference effects imply that the spectral searching of *mixtures* containing several components will not usually lead to unambiguous identification of the identity of the absorbers. However, the operation of searches in carefully selected wavenumber ranges may allow the spectral search to be focused upon the impurities themselves. For example, we recently subjected a sample of acetone vapour containing about 1% of an impurity to a spectral search. Searching between 4000–600 cm^{-1} using a vapour phase library identified (not surprisingly) the sample as being acetone. However, searching between 1080–980 cm^{-1} identified the sample as being methanol. Further investigation confirmed that the impurity in the acetone was indeed methanol, with the strong C–O vibration dominating the outcome of the search in the restricted wavenumber range.

The spectrum/structure display facility of search software is particularly useful when the compound is obscure. For example taking, at random, spectrum number 987. This is benomyl, a systemic fungicide, and it is useful to be able to observe on screen or to plot out as in Figure 7 when information of this kind is required.

5 MOLECULAR MODEL AND INFRARED SPECTRAL CORRELATIONS

Several software packages exist for building models suitable for easy display on a computer screen and which include the determination of energy of interaction within and between molecules. In many packages, models may be drawn in two-dimensions (2D) and converted into three-dimensions (3D) by automatic procedures. The backbone structure in terms of chains and rings is readily constructed and hydrogen atoms added automatically in terms of free available valencies. In this way models of molecules of varying complexity are readily constructed. Various thermodynamic values, including energies within molecules and energy changes following interactions between molecules, are determined from spectroscopic measurements. This section shows the potential for linking information from these two categories.

The display of structures using the various chemical modelling packages enables 3D images to be manipulated, emphasizing particular atoms and bonds by selection of suitable size and colour. Structures displayed on computer screen may be moved by rotation, translation or scaling up or down in size. Interacting molecules may be viewed from many different directions to determine favourable interactions or unfavourable steric hindrance.

The calculated potential energy of a molecule depends upon the masses, bond lengths and angles, and force field used. In the case of a diatomic molecule, the

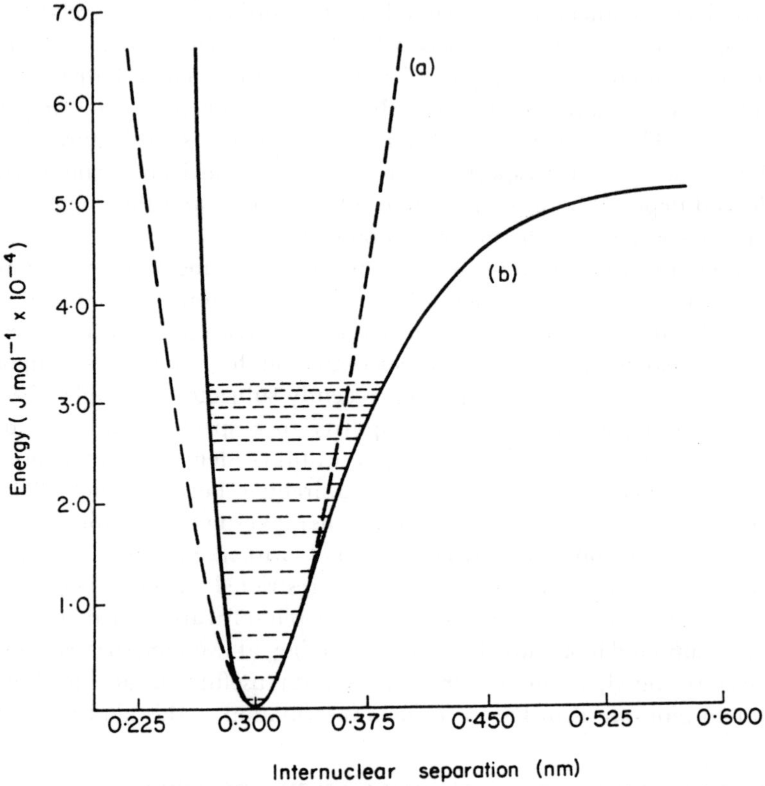

Figure 8 *The potential energy of a diatomic molecule (I_2):* (a) *harmonic oscillator,* (b) *anharmonic oscillator*
(Data computed and drawn by Dr A. J. Bowles.)

potential energy is a function of internuclear separation and follows a parabolic curve if the force field corresponds to a simple harmonic oscillator based on Hooke's law. This is a valid assumption for small changes in internuclear separation on which the law is based. If an anharmonic force field is chosen, the potential energy function is no longer a parabola but corresponds to the expected change in energy at very low and very high internuclear separations (Figure 8). These calculations follow wholly classical theories. Particular measurements of spectra provided evidence for breakdown of classical mechanical theory and emergence of quantum mechanics. Quantized energy levels are included in Figure 8 but, in general, modelling considerations are based on classical theory.

The value of the potential energy for a model of a molecule is clearly very much a consequence of the assumptions and theoretical basis of the calculations used. Molecular conformations each possess different energies and the rotation of a bond within a molecule is associated with a changing value for this potential energy. In the case of a polyatomic molecule the potential energy is a function of change of many internuclear separations and change of many bond angles. The rotation about any particular single bond leads to a continuously variable set of energy

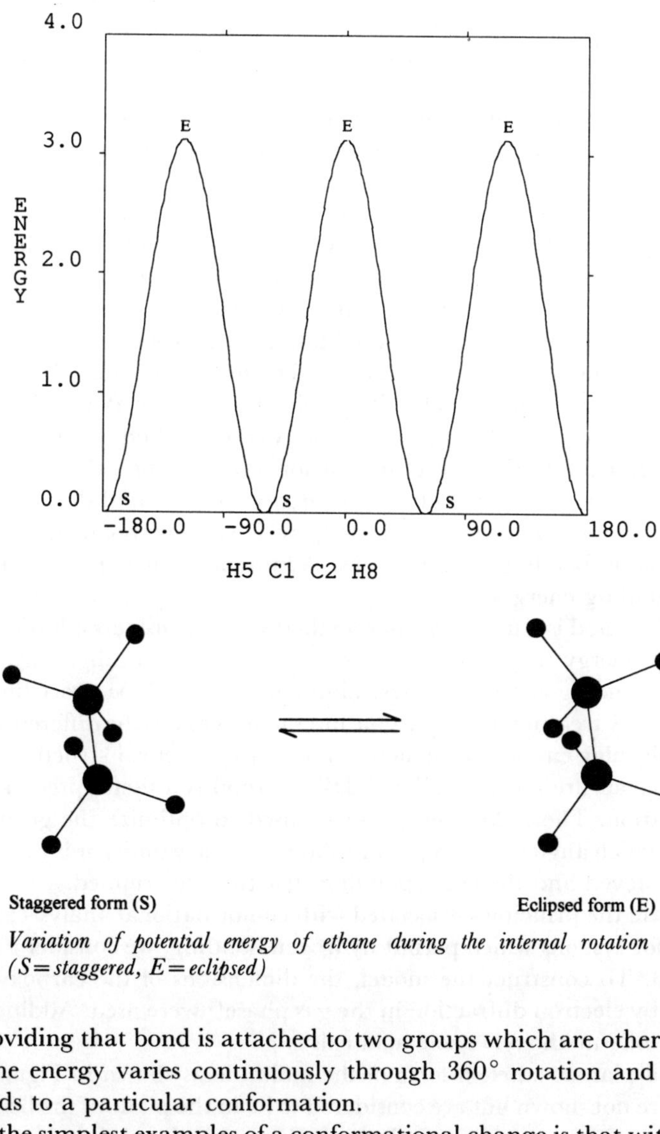

Figure 9 *Variation of potential energy of ethane during the internal rotation of C–C bond (S = staggered, E = eclipsed)*

values providing that bond is attached to two groups which are other than single atoms. The energy varies continuously through 360° rotation and each value corresponds to a particular conformation.

One of the simplest examples of a conformational change is that within ethane. At one time, it was believed that rotation about single bonds was unrestricted. In 1936, it was shown[13] that calculations of the entropy and enthalpy of ethane by the methods of statistical mechanics were inconsistent with results based on calorimetric values. These results were interpreted in terms of a barrier to rotation about the C–C single bond of about 3 kcal mol^{-1}. As one methyl group is rotated through 360° with respect to the other methyl group, the potential energy is found to follow a three-fold sine wave-type function. The potential energy function generated using the Chem-X package is shown in Figure 9 in which the barrier is determined

[13] J. D. Kemp and K. S. Pitzer, *J. Chem. Phys.*, 1936, **4**, 749.

using the Chem-X software. The minimum energy corresponds to a staggered conformation and the maximum energy corresponds to an eclipsed conformation. The structure of any particular conformation corresponding to a particular angle of rotation and potential energy may be determined at any point on the curve (by selection with the cursor prompt) and those corresponding to the staggered and eclipsed conformers are included in Figure 9.

The model of ethane built from the Chem-X modelling package uses standard bond lengths and angles for the C–C and C–H groups. If different values for these parameters are used, then the values of potential energy will change. Similarly, the choice of force field used for this calculation will also determine the quantitative values for energies within the same qualitative framework.

The energy of the system may be calculated by several methods. In the Chem-X package, energies are calculated by either using the Van der Waals (VDW) or the molecular mechanics (MM) options. The VDW method is an approximate method which includes the sum of the torsion strain energy, the non-bonded or polarization energy and the electrostatic energy. It is a rapid default method of calculating an approximate potential energy which ignores bond length and angle and out-of-plane bending contributions. Additionally, it is possible to include hydrogen bonding energy.

The MM method is a more rigorous method which considers additional factors to the VDW energy in terms of bond strain energy, bond angle strain energy, out-of-plane bending energy, and restraints energy. The MM procedure is based on a Hooke's law treatment of electronic interactions and is the difference between the intramolecular energy of the actual and a hypothetical structure where all structural aspects are idealized. The VDW method is a useful precursor to The MM calculation. The MM energy can be used to optimize the geometry of a structure in which angles and lengths are allowed to vary until a minimum value of energy is achieved and the corresponding structure determined.

To illustrate the principles associated with conformational analysis and energy calculation for systems which permit hydrogen bonding, we consider *o*-methoxybenzoic acid. To construct the model, the dimensions of the carboxylic group determined by electron diffraction in the gas phase[14] were used. Additionally, an *X*-ray study of *o*-methoxybenzoic acid exists.[15]

Consider a planar, labelled form of the molecule as shown in Figure 10. The hydrogens are not shown but are considered to lie in the plane of the benzene ring except for two of those on the methyl group. Four of the bonds in this structure are capable of rotation to provide conformational isomers. If the rotations are carried out in steps of 5° there will be 72 separate conformers for each bond and 72^4 total conformers, some of which are equivalent. There may not be a unique method for the determination of the lowest energy of all possible conformers but approximate methods may be used as follows.

Consider the rotation of bond 1 [O(10)–C(1)]. As the methoxy group rotates through 360°, the potential energy (E_p) changes as shown in Figure 11.

[14] J. L. Derissen, *J. Mol. Struct.*, 1971, **7**, 67.

[15] M. C. Etter, Z. U. Lipkowska, P. A. Fish, T. W. Panuanto and P. W. Baures, *J. Crystallogr. Spectrosc. Res.*, 1988, 18, **4**, 311.

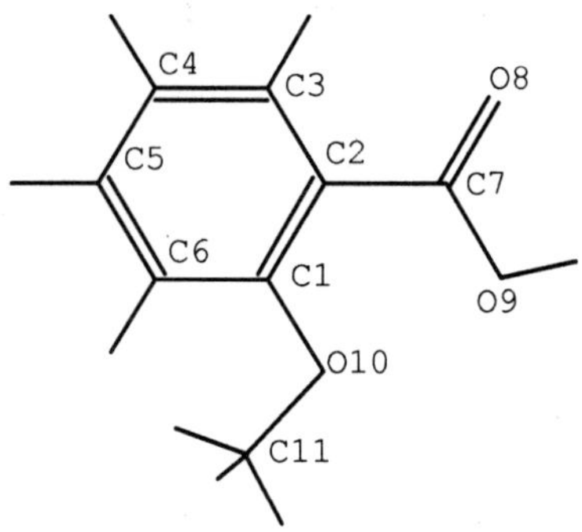

Figure 10 *Labelled atoms for* o-*methoxybenzoic acid (hydrogen omitted)*

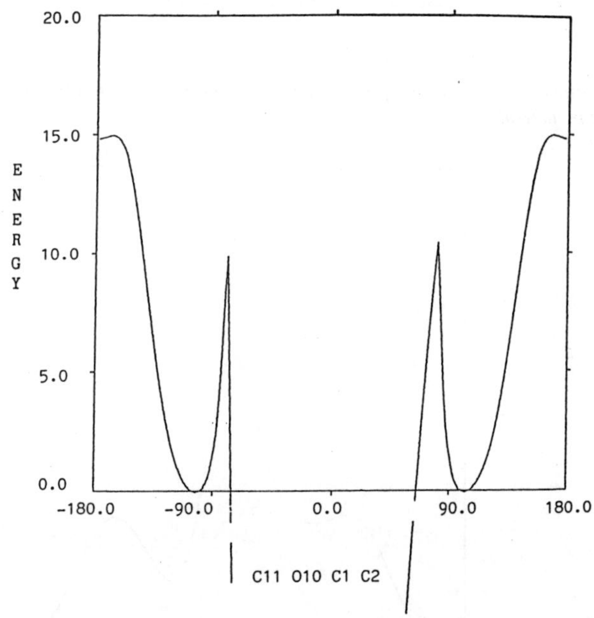

Figure 11 *Variation of potential energy of* o-*methoxybenzoic acid with rotation of* O(10)–*C*(1) *bond*

The two minima correspond to the methoxy group twisting 100° and −100° out of the plane. The diagram shows that as the methoxy group approaches the carboxylic acid group from either direction the E_p rises sharply until steric collision occurs when the graph drops off the scale. The conformers of least energy are identified as 57 and 17, corresponding to the double minima from the 72 possibilities.

Rotation of bond 2, C(7)–C(2) (the carboxylic group), illustrates the inter-relationship of 1D E_p plots as slices through a multi-dimensional E_p surface. For example, if the E_p diagram for bond 2 is calculated for a planar molecule,

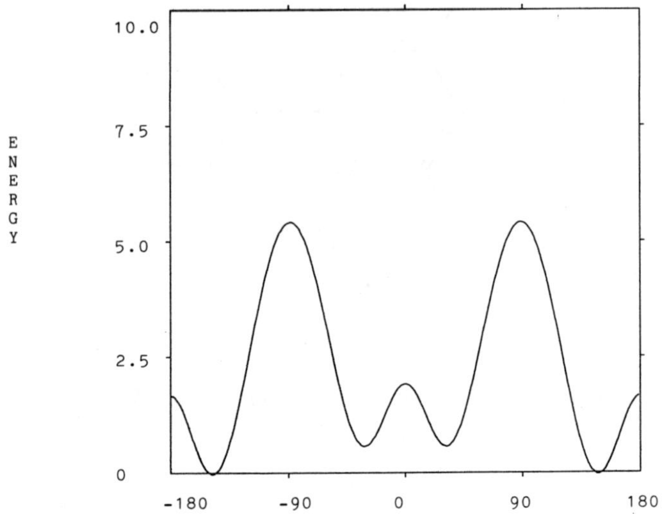

Figure 12 *Variations of potential energy of* o-*methoxybenzoic acid with rotation of* C(2)–C(7) *bond, planar molecule*

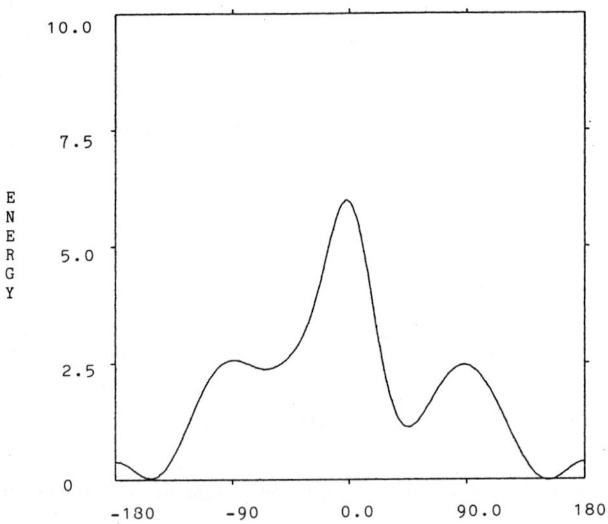

Figure 13 *Variation of potential energy of* o-*methoxybenzoic acid with rotation of* C(2)–C(7) *bond, non-planar molecule with minimum potential energy for* O(10)–C(1) *bond*

Figure 12 is obtained, which is a symmetric function, but if the diagram is calculated for the minimum E_p confomer (17) Figure 13 is obtained, which is an asymmetric function. In the latter case, the energy associated with the rotation of the carboxylic group clearly gives rise to different values according to whether the carboxylic group is on the same side or the opposite side as the methoxy group.

Rotation of bond 3, O(9)–C(7) (the hydroxy group), for the minimum value

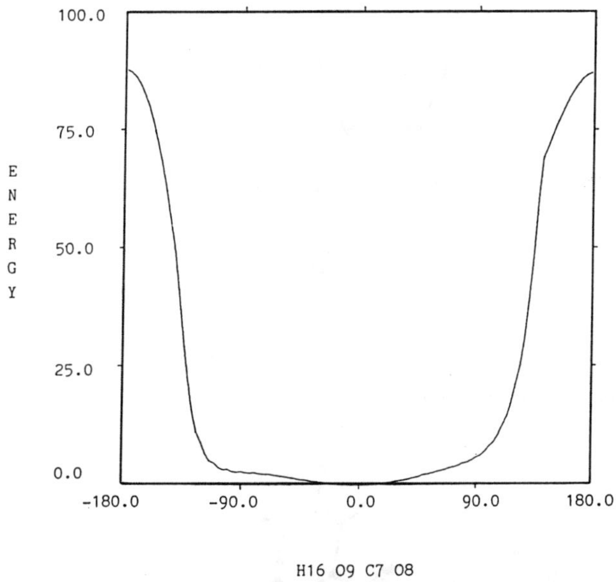

Figure 14 *Variation of potential energy of* o-*methoxybenzoic acid with rotation of O*(9)–*C*(7) *bond*

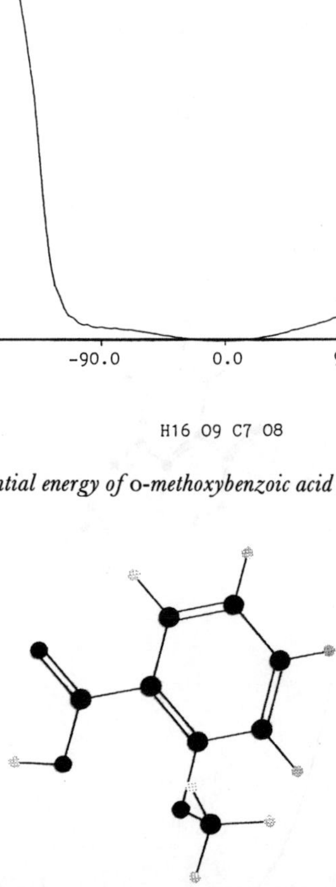

Figure 15 *Structure of* o-*methoxybenzoic acid for minimum energy in the absence of hydrogen bonding*

conformer for bond 1 and 2 rotations leads to an E_p diagram (Figure 14) in which the minimum energy is for the hydroxy group directed away from the methoxy group. The corresponding structure for this energy minimum is shown in Figure 15.

Experimental information provides an opportunity to refine the model. The IR spectrum of *o*-methoxybenzoic acid shows a prominent band at 3380 cm^{-1} assigned to the intramolecular hydrogen bonded form. This corresponds to structures shown in Figures 16 and 17.

The fourth bond which gives rise to rotational conformers is O(10) – C(11). Rotation of the methyl group above this bond provides a three-fold E_p diagram similar to that displayed in Figure 9. The package permits calculation of values for energies with and without hydrogen bonding that are based on assumptions which allow relative rather than absolute significance to be attributed, but serve as a

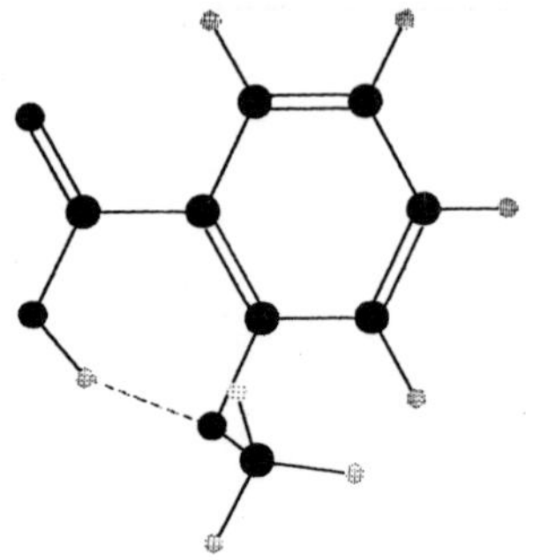

Figure 16 *Structure of* o-*methoxybenzoic acid for minimum energy following formation of an intramolecular hydrogen bond*

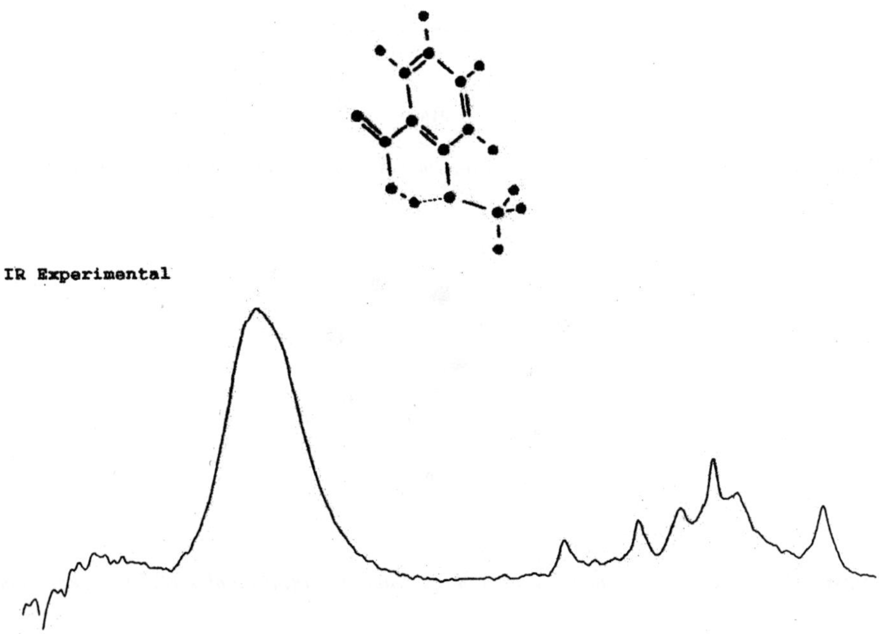

Figure 17 *Incorporation of infrared spectra of* o-*methoxybenzoic acid into Chem-X package* (3600–600 cm^{-1})

vehicle for further refinement. The stages in the calculation for the structure shown in Figure 16 are shown in Table 5 using the Chem-X software package (January 1994 release). This is termed a 'log' file and such files (for any set of operations) may be saved and printed as required. Note that in Table 5 the routines for the determination of the plot of energy diagrams and structures are omitted.

The Chem-X package permits modelling and spectroscopic information to be brought to the screen simultaneously. An example is shown in Figure 17. The

Table 5 *Log file for calculation on minimum VDW and MM energy of* o-*methoxybenzoic acid*

```
{Menu|Energy}
{Energy|(1): VDW__energy}
{*|VDW Minimise Tortions etc}
{Energy|Add: AUTO BONDS}
{*δOne Bond}
{Energy|Add Variable(s)}
{PICK ATOM|O10 ∧ molecule1}
{PICK ATOM|C1 ∧ molecule1}
{Energy|O10 C1}
{Energy|Calculate}
! VDW energy : 43.0104 kcals     Cycle   : 1
! VDW energy : 10.3558 kcals     Cycle   : 10
! % Convergence criterion reached
MEnergy|Add Variable(s)}
{PICK ATOM|C7 ˆ molecule1}
{PICK ATOM|C2 ˆ molecule1}
{Energy|C7 C2}
{Energy|Calculate}
! VDW energy : 10.29 kcals     Cycle   : 1
! VDW energy : 9.5358 kcals     Cycle   : 7
! % Convergence criterion reached
MEnergy|Add Variable(s)}
{PICK ATOM|O9 ˆ molecule1}
{PICK ATOM|C2 ˆmolecule1}
{Energy|O9 C7}
{Energy|Calculate}
! VDW energy : 9.5136 kcals     Cycle   : 1
! VDW energy : 9.5004 kcals     Cycle   : 4
! % Convergence criterion reached
{Energy|(1): VDW__MINIMISE}
{*|MM Optimise XYZ}
{Energy|Calculate}9999
! MM energy : 6.6104 kcals     Cycle   : 1
! MM energy : 3.441 kcals     Cycle   : 99
! % Energy convergence criterion reached – optimization terminated
MEnergy|(1): MM__OPTIMISE}
{*|VDW Energy}
{Energy|Calculate}
! VDW energy   : 7.3886 kcals
{Energy|OK}
```

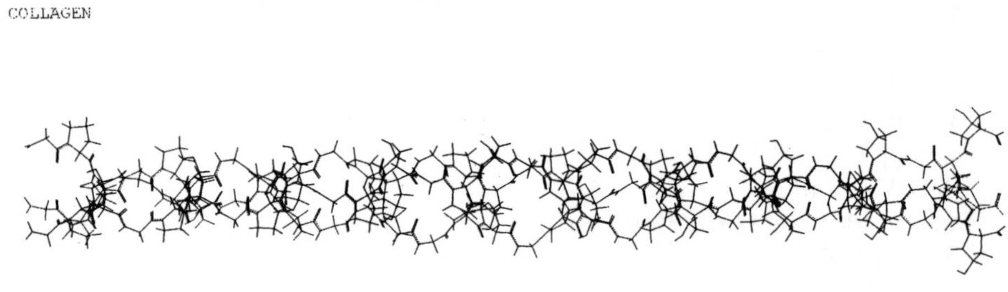

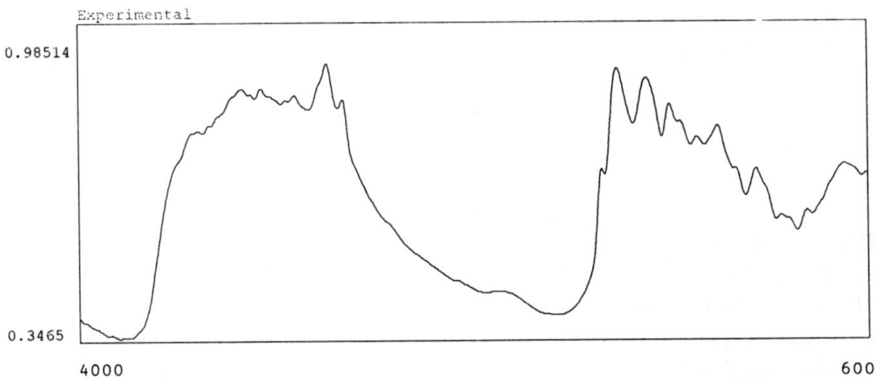

Figure 18 (a) *Molecular structure of collagen.* (b) *Infrared spectrum of collagen obtained by FTIR drift spectroscopy*

graphical information may be stored and transmitted as ASCII text files using standard protocols. The JCAMP and Chem-X CSSR text file associated with Figure 17 are shown in Tables 6 and 7, respectively.

These numerical data are a basis for further refinement and editing for iterative improvement of the relationship between modelling information and experimentally measured spectroscopic information.

An example of a biological macromolecule with hydrogen bonding features is collagen. The structure is well established[16] as three gelatin chains (each chain comprising repeating units of the amino acids proline, hydroxyproline and glycine) twisted into a triple helix held together by hydrogen bonds. The FTIR spectrum of collagen has been interpreted in terms of certain gross structural features.[17] Figure 18 shows a molecular model of collagen juxtaposed with an IR spectrum of a solid-state sample of collagen recorded by diffuse internal reflection (DRIFT) spectroscopy.

The importance of hydrogen bonding towards a better understanding of biological processes including drug action and DNA metabolism has been demonstrated by Wong.[18] A survey was carried out involving 156 women, 20 of

[16] P. Doty, 'Proteins', taken from 'The Molecular Basis of Life', An Introduction to Molecular Biology, readings from Scientific American, Freeman, 1968.

[17] K. J. Payne and A. Veis, *Biopolymers*, 1988, **27**, 1769.

[18] P. T. Wong, K. Wong, A. Caputo, A. Godwin and Rigas, *Proc. Natl. Acad. Sci. USA*, 1991, **88**, 10988.

Table 6 *JCAMP.DX file for the IR spectrum of* o-*methoxybenzoic acid for region of intramolecular hydrogen bond*

```
##Title=ortho-methoxybenzoic acid
##JCAMP-DX=4.24
##DATA TYPE=IR SPECTRUM
##ORIGIN=
##OWNER=
##DATE=94/02/16
##TIME=18:09:39
##SPECTROMETER/DATA SYSTEM=Perkin-Elmer 1720 IR
##X-INSTRUMENT PARAMETERS=
  DETECTOR=TGS
  APODIZATION=NORMAL
##RESOLUTION=4.000000
##DELTA-X=-1
##X-UNITS=1 cm-1
##Y-UNITS=ABSORBANCE
##X-FACTOR=1
##Y-FACTOR=1.19209289550781e-007
##FIRST-X=3600
##LAST-X=2800
##N-POINTS=801
##FIRST-Y=0.00487565994262694
##MAX-Y=0.364075
##MIN-Y=-0.005767
##XY-DATA=(X++(Y..Y))
3600.000+40900+47540+58440+54000+43280+45220+62760+82160+94840+90360+56140
3425.000+849720+868020+883560+899840+918960+943100+972100+1003340+1034300
3416.000+1063200+1090000+1115400+1142020+1172700+1208040+1247080+1289280
3408.000+1332960+1376500+1420580+1468800+1522320+1577540+1629760+1677940
3400.000+1725280+1774700+1825980+1878820+1932640+1986140+2037780+2089600
3392.000+2146000+2208740+2274980+2341580+2405580+2464440+2516840+2565720
3384.000+2614900+2665540+2715400+2761000+2803360+2844820+2883460+2913620
3376.000+2930200+2937820+2945760+2959340+2978640+3000880+3022940+3041160
3368.000+3051440+3054080+3049940+3041460+3031600+3022500+3014400+3006140
3360.000+2995660+2981920+2963140+2942380+2924840+2912760+2902580+2889760
3352.000+2872820+2853780+2833760+2811880+2786360+2758300+2731260+2706660
3344.000+2683060+2656780+2624920+2587660+2545480+2498680+2447040+2392220
3336.000+2340020+2295900+2258140+2220060+2176400+2128840+2080820+2034160
3328.000+1988660+1944840+1902480+1860540+1819020+1779680+1744000+1710280
3320.000+1674920+1636500+1597440+1560400+1526280+1493740+1461780+1430580
3312.000+1399260+1367940+1337540+1309040+1282720+1257980+1234380+1210760
3304.000+1186340+1161100+1136560+1113400+1089680+1063400+1035600+1010260
2800.000+437440
##END=
```

whom were suffering from cervical cancer or showed pre-cancerous conditions. Significant differences were found between the IR spectra of normal cervical cells and cancerous ones. Amongst the distinctive features the band produced by the phosphodiester group in DNA was broader in the cancerous cells and one of the component bands intensified in malignant cells where Wong believes that a larger number of phosphodiester groups become hydrogen bonded to water. It is hoped that these discoveries will lead to improvements over the traditional screening methods for early detection and treatment of potential cervical cancer in women.

TABLE 7 *CSSR Text file*

ortho-*Methoxybenzoic Acid*													
1 C(1)	−5.77798	−2.50276	−0.49016	2	6	10	0	0	0	0	0	0.089	1
2 C(2)	−5.20829	−1.38847	−1.08116	1	3	7	0	0	0	0	0	0.076	1
3 C(3)	−5.97753	−0.26345	−1.32210	2	4	12	0	0	0	0	0	−0.046	1
4 C(4)	−7.31649	−0.25268	−0.97205	3	5	13	0	0	0	0	0	−0.061	1
5 C(5)	−7.88623	−1.36695	−0.38105	4	6	14	0	0	0	0	0	−0.059	1
6 C(6)	−7.11703	−2.49199	−0.14009	5	1	15	0	0	0	0	0	−0.031	1
7 C(7)	−3.75130	−1.40015	−1.46208	2	8	9	0	0	0	0	0	0.294	1
8 C(8)	−3.00689	−2.36324	−1.27854	7	0	0	0	0	0	0	0	−0.259	1
9 O(9)	−3.34925	−0.35711	−2.02008	7	16	0	0	0	0	0	0	−0.327	1
10 O(10)	−5.02485	−3.60420	−0.25427	1	11	0	0	0	0	0	0	−0.347	1
11 C(11)	−5.85893	−4.56520	−0.34612	10	17	18	19	0	0	0	0	0.043	1
12 H(12)	−5.54530	−0.58192	−1.77048	3	0	0	0	0	0	0	0	0.063	1
13 H(13)	−7.90008	−0.60085	−1.15485	4	0	0	0	0	0	0	0	0.062	1
14 H(14)	−8.90207	−1.35877	−0.11548	5	0	0	0	0	0	0	0	0.062	1
15 H(15)	−7.54927	−3.33735	−0.30829	6	0	0	0	0	0	0	0	0.064	1
16 H(16)	−2.34965	−0.31544	−2.26154	9	0	0	0	0	0	0	0	0.221	1
17 H(17)	−5.29737	−5.46429	−0.55276	11	0	0	0	0	0	0	0	0.053	1
18 H(18)	−6.25808	−4.18049	−1.27299	11	0	0	0	0	0	0	0	0.053	1
19 H(19)	−6.67816	−4.80772	−0.31452	11	0	0	0	0	0	0	0	0.053	1

6 CONCLUSION

The computer networking of spectroscopic instruments provides many opportunities for solving problems of analytical, physical, biological and medical importance. Computer methods cannot improve the basic data, in fact these inevitably corrupt data to a greater or lesser extent. However, much data; is incomprehensible until turned into information which can be understood and correlated with other information. It is in these roles that computers have become invaluable. It is also well realized that much of this computer power is now available on low priced portables and laptops. The juxtaposition of spectroscopic information and molecular model/energy value information on screen under real-time control has applications both to research and to education.

The authors are grateful to Rahima Dilshad, Ray Hancock, Brian Minty and Jason Price for technical support and data collection and to Aston Scientific and Chemical Design Ltd for advice on software development.

CHAPTER 6

Multivariate Methods for Automated Spectrum Interpretation

H.J. LUINGE

1 INTRODUCTION

Automation of spectrum interpretation has been a goal of spectroscopists for many years. Early attempts to have machines compare spectra with collections of reference data even date back to the early 1950s. Since then enormous progress has been made with respect to the speed of search for the correct reference and the amount of data digitally available. Still, however, library search systems suffer from the disadvantage that in order to solve the structure of an unknown compound unambiguously the corresponding reference spectrum has to be available. When this is not the case, the unknown may be structurally similar to the references with the most similar spectra. Furthermore, large libraries are required when divers problems are to be solved. In an attempt to overcome these problems classification systems have been developed, capable of assigning the spectrum of an unknown compound to one or more structural classes. Even when no correct reference spectrum is available, detailed structural information can be obtained this way. The classifications are based on decision rules obtained from mathematical processing of example cases. The amount of data to be assessed during the classification process is limited, which greatly enhances the speed of these systems.

Several approaches have been undertaken to classify spectra of unknown samples according to the presence and/or absence of structural fragments. These approaches can be divided mainly into rule-based and multivariate methods. The rule-based methods usually apply spectrum–structure correlations in terms of rules (*e.g.* if there is a strong band around 1700 cm^{-1}, then the compound is likely to contain a carbonyl group). Several automated systems based on this approach have been developed. Here, we will focus on another group of methods that apply mathematical/statistical techniques to the data in order to obtain structural information. Although these classification methods can be applied to any kind of multivariate data, attention will be focused on classification methods for

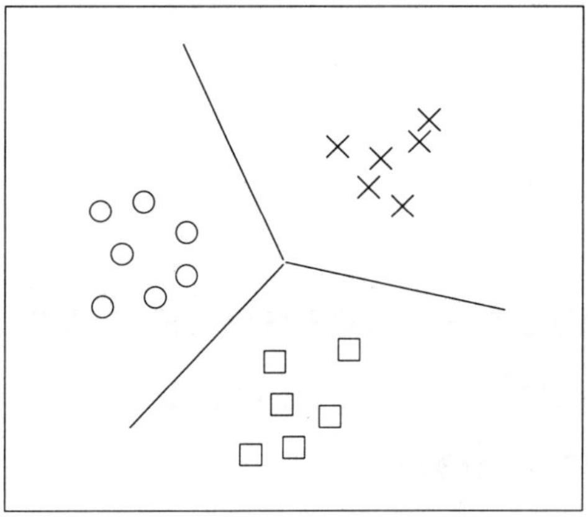

Figure 1 *Example of a two dimensional three-class problem with three decision lines*

vibrational spectra. A review on automated interpretation of vibrational spectra in general has appeared elsewhere.[1]

2 THEORY

Spectra can be regarded as objects in a multi-dimensional pattern space. The axes in this space represent the different wavenumbers and the coordinates the corresponding intensities. The classification methods discussed in this chapter attempt to separate patterns of one class from patterns of other classes by defining one or more decision surfaces (non-parametric methods) as is illustrated in Figure 1 or by developing statistical models for each class (parametric methods).

The above-mentioned methods are called supervised as the classification rules are determined from *a priori* imposed classes. Alternatively, unsupervised methods can be applied. In that case no knowledge of the presence of different classes is used in the analysis. Mathematical techniques are used to explore the pattern space and possibly find clusters corresponding to spectra with a high degree of similarity. Assuming that similar spectra correspond with similar structures, conclusions can be drawn on the presence of structural classes. It should be noted that these classes are based purely on spectral similarities. Hence, the classes thus obtained may differ from conventionally used classification schemes. Subsequently, supervised methods can be used to obtain the corresponding classification rules.

2.1 Supervised Techniques

Supervised classification methods can be subdivided into parametric and non-parametric methods. In the former case a statistical distribution of the objects in pattern space is assumed. When non-parametric techniques are applied no assumptions as to the distributions are made.

[1] H.J. Luinge, *Vib. Spectrosc.*, 1990, **1**, 3.

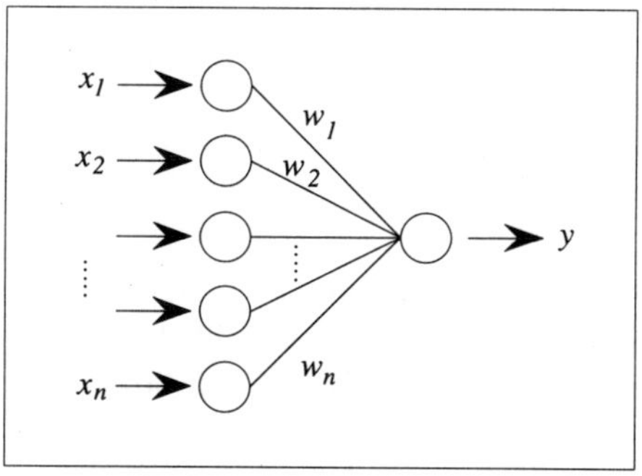

Figure 2 *Schematic representation of a perceptron*

2.1.1 Non-parametric Methods. Non-parametric classification methods attempt to develop decision surfaces between classes. The orientation of these surfaces is determined during the training phase in which the positions of known patterns are investigated and compared. Often different orientations of the decision surfaces can be found to separate classes, and the surface obtained in each particular case depends on the training algorithm used.

A spectrum containing n data points $x_1, x_2, \ldots x_n$ can be regarded as a vector $\boldsymbol{x}$ in n-dimensional space. A binary pattern classifier divides this space into two subspaces separated by a decision surface. The orientation of a decision surface is commonly described by a number of weights. When dealing with linear separation methods these weights correspond with the components of a vector perpendicular to the decision plane. The decision surface can then be defined by a linear discriminant function of the form:

$$s = f(\boldsymbol{x}) = w_0 + w_1 x_1 + w_2 x_2 + \ldots + w_n x_n,$$

where $f(\boldsymbol{x}) \geq 0$ corresponds to one class and $f(\boldsymbol{x}) < 0$ to the other. Thus, this binary classifier assigns patterns to either of two classes depending on which side of the decision surface they fall. The linear combination can be regarded as the dot product between the spectrum vector $\boldsymbol{x}$ and the weight vector $\boldsymbol{w}$. The weight vector is perpendicular to the decision plane ($s = \boldsymbol{w} \cdot \boldsymbol{x} = 0$).

A graphical representation is given in Figure 2, where the concept of a *perceptron* is shown. In Figure 2 inputs x_i (*e.g.* spectral intensities) are coming from the left and are multiplied by weights w_i assigned to the lines connecting the input units to the output unit. The output value y usually is a binary number and corresponds with the presence or absence of a particular class. The predicted value of y is obtained by summation of the products $x_i w_i$ (*i.e.* the dot product between input vector $\boldsymbol{x}$ and weight vector $\boldsymbol{w}$). This weighted sum is compared with a threshold value (usually 0) and the input pattern is classified accordingly. The weights w_i are determined during a training phase in which known patterns and their outputs are presented to the classifier.

Training of the classifier is achieved by adapting the weight vector $\boldsymbol{w}$. Initially, all weights are set to any convenient value. Next, a spectral pattern is presented and the value of s is calculated and compared with the desired value. If the response is correct, no changes are made. Otherwise, a new weight vector $\boldsymbol{w}'$ is calculated as:

$$\boldsymbol{w}' = \boldsymbol{w} + c \cdot \boldsymbol{x},$$

where c is the correction factor. This factor is determined in such a way that the spectral pattern $\boldsymbol{x}$ will be on the correct side of the decision plane. One way of achieving this is to reverse the sign of the dot product $\boldsymbol{w} \cdot \boldsymbol{x}$.

$$-\boldsymbol{w} \cdot \boldsymbol{x} = \boldsymbol{w}' \cdot \boldsymbol{x} = (\boldsymbol{w} + c \cdot \boldsymbol{x}) \cdot \boldsymbol{x} = \boldsymbol{w} \cdot \boldsymbol{x} + c \cdot \boldsymbol{x} \cdot \boldsymbol{x} = s + c \cdot \boldsymbol{x} \cdot \boldsymbol{x}.$$

Hence, it follows that:

$$c = -2 \cdot \boldsymbol{w} \cdot \boldsymbol{x} / \boldsymbol{x} \cdot \boldsymbol{x}.$$

This method is called error correction feedback training. When the classes of patterns are truly linearly separable, repeated presentation will ultimately yield a weight vector which gives correct classifications for all patterns.

A so-called dead zone z can be added to the decision surface by calculating:

$$c = 2 \cdot (\pm z - \boldsymbol{w} \cdot \boldsymbol{x}) / \boldsymbol{x} \cdot \boldsymbol{x},$$

where the sign of z depends on the sense of the error. Patterns for which the dot product with the weight vector lies between $\pm z$ are regarded as not classifiable.

The orientation of the decision surface is dependent on the initial weights. Hence, it is common practice to set all weights initially to negative values (*e.g.* -0.1) and repeat the entire training process with all weights set to positive values (*e.g.* $+0.1$). In this way two decision surfaces are obtained. The two weight vectors can be used for classification of unknowns. Patterns between the two decision planes can be regarded as unknowns. Kowalski *et al.*[2] applied this method to the classification of IR spectra and demonstrated that the reliability of predictions increases, however, at the cost of fewer patterns classified. Alternatively, starting values for the weight vector can be calculated by subtracting the average non-class spectrum from the average class spectrum. Usually, this results in faster convergence of the iteration process.

Instead of iteratively producing weight vectors by error feedback correction, average class spectra can be used as final weight vectors. This representation has the advantage of showing all of the bands characteristic of the class at hand preserving relative intensities. A variation on this approach is the use of muted average spectra: the average non-class spectrum is subtracted from the average class spectrum thus reducing the noise level, however at the cost of reducing the intensity of common bands.

[2] B. R. Kowalski, P. C. Jurs, T. L. Isenhour and C. N. Reilley, *Anal. Chem.*, 1969, **41**, 1945.

By definition, linear discriminant analysis (LDA) methods compute the orientation of a linear separating surface. In the case where two classes cannot be separated linearly then a non-linear decision surface has to be calculated. This can be achieved by including higher order terms in the discriminant function (*e.g.* quadratic). Another extension of LDA allowing the use of non-linear surfaces is piecewise linear discriminant analysis, which uses multiple linear discriminants to form a piecewise approximation of a non-linear surface. Kaltenbach and Small[3] recently applied this method in combination with ridge regression to the classification of FTIR remote sensing data.

Generally, many classifiers can be obtained that are capable of separating two classes. The predictive ability of the classifier, however, may vary when different training algorithms are used. Nilsson[4] described a variety of alternative discriminant functions.

An alternative approach is to calculate the distance $|\boldsymbol{x}-\boldsymbol{w}|$ instead of the dot product between a pattern $\boldsymbol{x}$ and a weight vector $\boldsymbol{w}$. Minimum distance classifiers are appropriate in cases where a class can be described by a weight vector around which all patterns in that class tend to cluster. Woodruff *et al.*[5] compared the dot product and distance methods for classification of binary IR data. It appeared that for binary data the distance between pattern and weight vector was a better classifier than the dot product.

2.1.2 Artifical Neural Networks. A field of major interest in current chemometric research is the appliction of artificial neural networks (ANNs) to a variety of chemical problems. An overview has recently been given by Zupan and Gasteiger.[6] ANNs were originally inspired by the human brain, where networks of nerve cells interconnected by synapses learn by processing information from example cases.[7] ANNs are similar only to the extent that they consist of networks of units (neurons), interconnected by weights that are trained by processing example data. A schematic representation of an ANN is given in Figure 3.

Usually an ANN consists of three or more layers of neurons: an input layer, an output layer and one or more hidden layers. Each of the neurons in the network receives signals from all neurons in the previous layer (or directly from the input source). These signals are weighted according to the weights assigned to each of the interconnections. A (usually non-linear) transfer function is applied to the sum of the weighted signals resulting in an output signal for the neuron. This output is transferred to the next layer of neurons (or directed towards the output channel). This process is schematically depicted in Figure 4. In principal, the ANN approach is merely an extension of the work on previously described discriminant classifiers.

The most common transfer function used is the sigmoid function:

[3] T. F. Kaltenbach and G. W. Small, *Anal. Chim. Acta*, 1993, **279**, 309.
[4] N. J. Nilsson, 'Learning Machines', McGraw-Hill, New York, 1965.
[5] H. B. Woodruff, S. R. Lowry and T. L. Isenhour, *Appl. Spectrosc.*, 1975, **29**, 226.
[6] J. Zupan and J. Gasteiger, *Anal. Chim. Acta*, 1991, **248**, 1.
[7] J. Dayhoff, 'Neural Network Architectures', Van Nostrand Reinhold, New York, 1990.

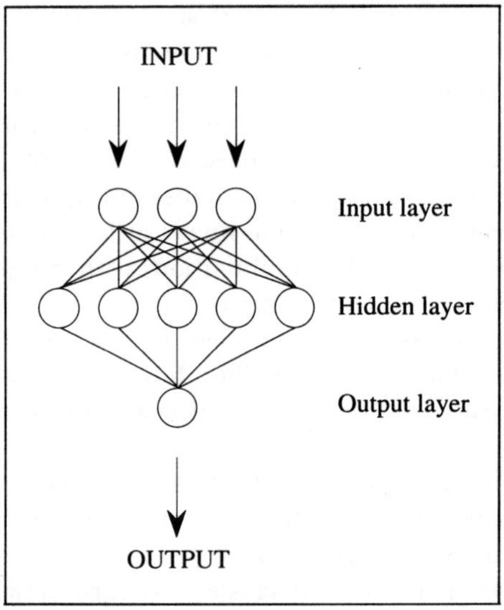

Figure 3 *Schematic representation of an artificial neural network*

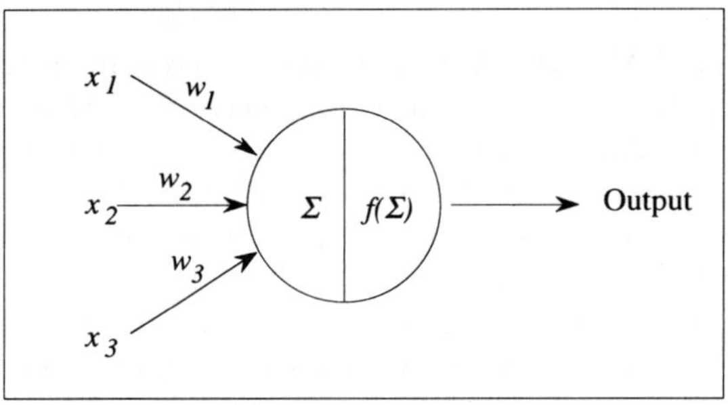

Figure 4 *Schematic representation of the processes taking place within a single artificial neuron*

$$f(net) = \frac{1}{1 + \exp\left(\frac{-net}{T} + \theta\right)}$$

$$net = \sum_i x_i w_i$$

The sigmoid function has an s-shaped form. A plot of the output of a neuron as a function of its net input is given in Figure 5. The position of the inflection point and the slope are determined by the so-called *bias* value θ and the *temperature* T.

After presentation of a pattern to the neural network and subsequent forward propagation yielding output values, an error value is computed by comparing the output with the target value. The error value is calculated as:

$$\delta_j = (t_j - o_j) \cdot f'(S_j)$$

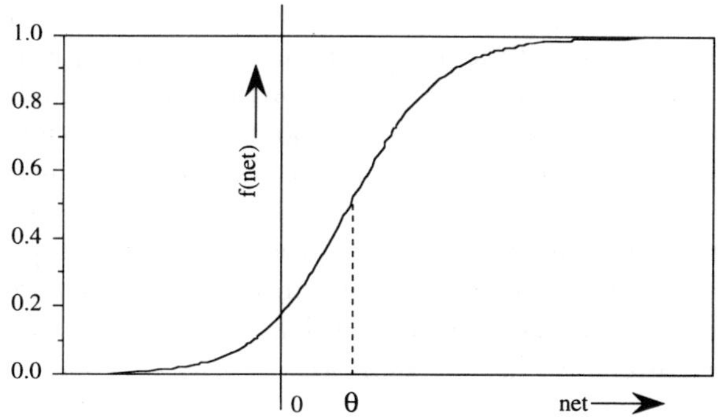

Figure 5 *Sigmoid function*

where

t_j = the target value for unit j
o_j = the output value for unit j
$f'(x)$ = the derivative of the transfer function
S_j = the weighted sum of inputs to unit j.

The difference $(t_j - o_j)$ reflects the amount of error, whereas the derivative f' scales the error towards a stronger correction when the sum S_j is near the rapid ascent in the transfer function.

For units in the hidden layer no target value is available. For these units error values are calculated according to:

$$\delta_j = [\Sigma\ \delta_k w_{kj}] \cdot f'(S_j)$$

Here, a weighted sum is taken of the errors of all units k that receive input from unit j. Clearly, for this reason the training method is called the error backpropagation algorithm.

The connection weight between unit i and j is adjusted according to:

$$\Delta w_{ij} = \eta \cdot \delta_j \cdot o_i,$$

where

Δw_{ij} = the adjustment of the weight from unit i to unit j
η = the learning rate
δ_j = the error value corresponding with unit j
o_i = the output from unit i.

This equation is known as the generalized delta rule.[8] The size of the weight adjustment is proportional to the error value of the target unit. A large error thus

[8] D. E. Rumelhart and J. L. McClelland, 'Parallel Distributed Processing', MIT Press, Cambridge, 1986, vols. 1 and 2.

results in a large adjustment of the incoming weights. The adjustment is also proportional to the output of the originating unit. If this output is large, a large weight adjustment will be made. The proportionality constant η reflects the rate of learning and is usually chosen between 0.25 and 0.75. A low value will result in slow-learning behaviour of the network, whereas a too high value may give rise to oscillations.

In order to speed up the training process several adaptations to the backpropagation algorithm have been developed. The most commonly used one incorporates a momentum factor m in the weight adaptation equation thus taking the previous weight change into account as:

$$\Delta w_{ij}(t+1) = \eta \cdot \delta_j \cdot o_i + m \cdot \Delta w_{ij}(t).$$

The quantity $t+1$ indicates the $(t+1)$th iteration.

During the learning phase each pattern of the training set is presented to the network and the weights are adapted according to the predicted output values. This process is repeated until some convergence criterion is obeyed. Commonly the root-mean-square (RMS) error is used for this purpose, computed as:

$$RMS = \sqrt{\frac{\Sigma_p \Sigma_j (t_{jp} - o_{jp})^2}{n_p n_o}}$$

where

n_p = number of patterns in the training set
n_o = number of units in the output layer.

Several attempts have been undertaken to apply ANNs for the interpretation of IR spectra. Robb and Munk[9] investigated the use of a simple linear neural network for this purpose. This network consists of an input and an output layer only (no hidden units) and no transfer function is used. Hence, the outputs are direct linear combinations of the inputs. This simplicity of network topology has some fundamental limitations and may produce less favourable results than more complex models. Actually, the network architecture used is equivalent to that of a linear discriminant function, but the training algorithm differs. Owing to this network simplicity the relationship between input and output values is readily interpretable. The training set consists of 6695 spectra of widely differing structure. Each spectrum is transformed into a peak table. Subsequently an input vector is constructed by dividing the range 4000–400 cm^{-1} into 640 intervals each of which is assigned to one input unit. Only when an interval contains a peak is a non-zero value assigned to the corresponding input. This value is determined by subtracting the % transmission of the peak from 100 and dividing the result by 100. This way input values between 0 and 1 are obtained. Alternatively, the number of inputs units is reduced to 256 and a more detailed sampling is applied for the low wavenumber region by decreasing the width of the intervals linearly

[9] E. W. Robb and M. E. Munk, *Mikrochim. Acta*, 1990, **I**, 131.

with decreasing wavenumber from *ca.* 20.8–6.6 cm^{-1}. The output vectors are binary vectors determined by the presence or absence of any of 128 different functional groups. These are general (*e.g.* aromatic) or more specific (*e.g.* Ar–O–CH_3). The results obtained are summarized using an A_{50}-value, which is defined as the reliability for a positive identification (number of correct positives divided by the total number of positive identifications) at the 50% recovery threshold. This threshold represents the output value for which 50% of the samples with a particular functionality is positively identified. The trained network was tested on a set of 541 spectra and 24 functional groups. In approximately half of the cases an assertion about the presence or absence of a fragment was made. Four assertion levels were used: definitely absent, possibly present, probably present and definitely present. The overall accuracy of assertions at these levels was 99.5, 48.0, 69.2 and 87.6%, respectively. Easily recognizable fragments are C=O, triple bond/cumulene, nitrile and phenyl (C_6H_5); more difficult ones for instance are C–halogen, C–O and ether.

In a second paper Munk *et al.*[10] describe the application of a three-layered network with sigmoid transfer functions. They concluded that a substantially better performance was obtained than with the simple linear architecture.

Meyer and Weigelt[11] applied a neural network to low resolution IR spectra consisting of 128 data points in the region 4000–400 cm^{-1}. The network contained one hidden layer of 40 units and was trained to recognize 32 structural fragments. Based on 100 training and 50 test spectra, the percentages of false positives and negatives varied from 4.3–5.1% and 17.4–23.8%, respectively at different prediction levels. In order to obtain information on relevant spectral regions for a particular fragment functional group spectra were calculated with an iterative algorithm yielding inputs corresponding with a desired output pattern.

In a subsequent paper[12] Meyer *et al.* further reduced the number of spectral data points by using 4–10 principal component scores. The network contained a hidden layer with 30 units and was trained to recognize 13 fragments. A total of 700 spectra was used for training, whereas 348 spectra were used for testing the trained network. They demonstrated that the ability to recognize the training data improves with the number of scores used. For the test data a number between 4 and 10 scores was in most cases sufficient to classify unknowns into the 13 different classes. The percentage false positives ranged from 0% (sulfo group, 4 scores) to 30% (aliphatic C–H group, 10 scores), whereas the false negatives varied between 0% (carboxylic acid, 6 scores) and 64% (aldehyde, 8 scores).

Ricard *et al.*[13] investigated three types of network architectures: (*a*) no hidden layer and a single output, (*b*) a hidden layer of 33 nodes and a single output, and (*c*) a hidden layer of 33 nodes and 12 outputs. The number of input nodes was 284 in all cases (4000–600 cm^{-1}) and the number of training and test spectra 212 and 236, respectively. They suggest the use of several performance indices as it is easy to obtain very optimistic evaluations of the network's performance from certain

[10] M. E. Munk, M. S. Madison and E. W. Robb, *Mikrochim. Acta*, 1991, **II**, 505.
[11] M. Meyer and T. Weigelt, *Anal. Chim. Acta*, 1992, **265**, 183.
[12] M. Meyer, K. Meyer and H. Hobert, *Anal. Chim. Acta*, 1993, **282**, 407.
[13] D. Ricard, C. Cachet, D. Cabrol-Bass and T. P. Forrest, *J. Chem. Inf. Comput. Sci.*, 1993, **33**, 202.

indices, particularly for highly and thinly populated categories. The results show that in almost all cases the multilayer network trained to recognize one group at a time performs better than the other networks. However, in most cases the performance of the network without a hidden layer is almost as good as that of the one with a hidden layer, although for some fragments (*e.g.* C–CHO) the latter improves results drastically.

Smits *et al.*[14,15] propose an architecture of modular neural networks at hierarchical levels. They illustrate the approach with a system containing a network for the detection of carbonyl and hydroxy groups (2 output units) and two networks at a lower level for specification of the type of carbonyl (6 outputs) and hydroxy (4 outputs) present. Up to 2904 training and 650 test spectra are used, each reduced to 268 data points. The number of hidden units is set to 30. It was found that the hierarchical architecture proves to perform better than a conventional flat one. The overall performance was found to be similar to that of a genuine expert.

Recently, we have demonstrated[16,17] the successful application of ANNs for detailed tasks in the field of spectral image recognition. Attention was focused on three different objectives: (*i*) recognition of global spectral profiles for fast initial classification (*e.g.* organic, inorganic and aromatic); (*ii*) recognition of characteristic local spectral patterns (*e.g.* O–P=S in organophosphorous pesticides); (*iii*) recognition of bandshapes (*e.g.* –O–H, >N–H, ≡C–H. In all cases ANNs could be used with a high degree of success.

2.1.2 Parametric Methods. Parametric methods assume that the classes of patterns under investigation can be adequately described by statistical distributions. These distributions are to be derived from a training set. Subsequently, probabilities for the membership of each class can be calculated.

An example of a parametric method is given by Lowry *et al.*[18] They apply a maximum likelihood classifier to binary IR spectra. By calculating conditional probabilities (p_{ij}) for the presence and absence of a peak ($x_i = \pm 1$) given a particular class (c_j) and multiplying these probabilities for all spectral features, a discriminant function is obtained. By taking the logarithm of this function, one can derive that a simple linear classifier can be calculated with a weight vector:

$$w_{ij} = \log[p_{ij}/(1-p_{ij})].$$

Woodruff *et al.*[19] compared this maximum likelihood classifier with non-parametric methods based on the dot product and several distance measures, and found that the parametric approach gave superior results.

Statistical linear discriminant analysis (SLDA)[20] calculates decision surfaces

14 J. R. M. Smits, P. Schoenmakers, A. Stehmann, F. Sijstermans and G. Kateman, *Chemom. Intell. Lab. Syst.*, 1993, **18**, 27.
15 Q. C. van Est, P. J. Schoenmakers, J. R. M. Smits and W. P. M. Nijssen, *Vib. Spectrosc.*, 1993, **4**, 263.
16 H. J. Luinge, J. H. van der Maas and T. Visser, 'Book of abstracts: Euroanalysis VIII', Edinburgh, 1993, PR11.
17 T. Visser, H. J. Luinge and J. H. van der Maas, *Anal. Chim. Acta*, 1994, **296**, 141.
18 S. R. Lowry, H. B. Woodruff, G. L. Ritter and T. L. Isenhour, *Anal. Chem.*, 1975, **47**, 1126.
19 H. B. Woodruff, G. L. Ritter, S. R. Lowry and T. L. Isenhour, *Appl. Spectrosc.*, 1976, **30**, 213.

between classes based on the properties of the probability distributions of the classes. The decision plane is oriented in such a way that the ratio between inter- and intra-class variance is maximized. This approach only works well if the statistical assumptions are met, *i.e.* when the variance–covariance matrices of the classes are equal.

2.1.3 Modelling Methods. Categories can be described by principal component models determined separately from the training data for each category. When the data are arranged in an $m \times n$ matrix $\boldsymbol{A}$ with m the number of variables (wavenumbers) and n the number of objects (spectra), the covariance matrix can be written as $\boldsymbol{C} = \boldsymbol{A} \cdot \boldsymbol{A}^T$. The eigenvectors of this matrix are linear combinations of the original variables and the eigenvalues are related to the amount of variance accounted for by the corresponding eigenvector. A projection of the objects onto the most important eigenvectors allows visual inspection of the multi-dimensional pattern space in two or three dimensions. By representing each object using the coordinates in this space of reduced dimensionality, fewer data points are to be handled in the classification analysis. Once principal component models have been obtained for each of the classes, the validation data are fitted using these models and assigned to the most probable class. An example of a method applying this technique is called SIMCA.[21]

Domokos *et al.*[22] applied the SIMCA method to the classification of carboxy, alcohol, ester, amine and aromatic compounds. Depending on the scaling operation performed on the original data percentages of correctly classified samples ranged from 72% (binary scaling) to 93% (auto-scaling). Multi-category linear separation appeared to give superior results, however (91–96%).

2.2 Unsupervised Techniques

Classification of spectra is usually based on pre-defined classes corresponding to structural fragments. In other words, structural similarities are used to develop classification functions for the spectra. Which functionalities are best suited for use in classification depends on the separability of the spectra in pattern space. Alternatively, one can define classes based on spectral similarities. In pattern space these classes will appear as clusters of objects. Clearly, in order to be of use for interpretation these classes have to be related to structural features. In order to obtain classes of spectrally similar objects unsupervised pattern recognition methods can be applied. Two different approaches towards the perception of clusters will be discussed: conventional cluster analysis and Kohonen networks.

2.2.1 Cluster Analysis. Detection of clusters in n-dimensional space ($n > 3$) is only achievable using mathematical algorithms. There are two approaches to produce clusters: a hierarchical and a partitioned one. Hierarchical clusters contain other clusters which contain other clusters and so on. Hence, these can be represented by

[20] D. L. Massart, B. G. M. Vandeginste, S. N. Deming, Y. Michotte and L. Kaufman, 'Chemometrics: a Textbook', Elsevier, Amsterdam, 1988.

[21] S. Wold, *Patt. Recogn.*, 1970, **8**, 127.

[22] L. Domokos, I. Frank, G. Matolcsy and G. Jalsovszky, *Anal. Chim. Acta*, 1983, **154**, 181.

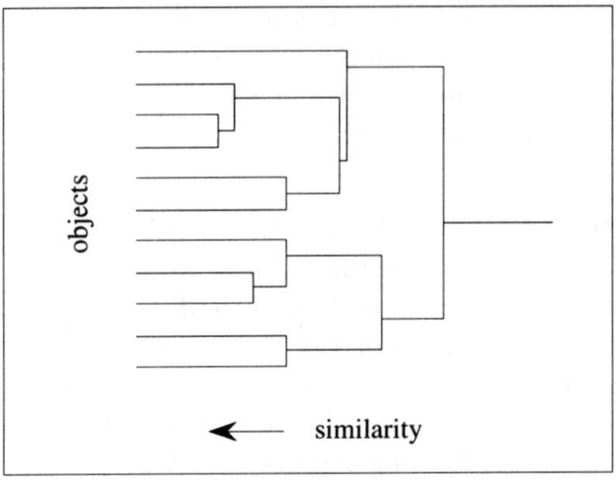

Figure 6 *Tree obtained from hierarchical clustering of IR spectra*

a tree as illustrated in Figure 6. Partitioned clusters contain no other clusters.

All cluster analysis methods are based on distances between patterns in pattern space. Various methods can be used to compute the distance of one object or cluster from another. Clusters are obtained by merging objects or clusters with the closest of the remaining objects or clusters. Several merging or linkage methods are available to produce hierarchical clusters. One method uses the distance between two objects or clusters as the distance between the two closest members of those clusters. This method is often called *single linkage* clustering.[23] It tends to produce long, stringy clusters. The *complete linkage* method uses the most distant pair of objects in two clusters to compute between-cluster distances, thus producing compact, globular clusters. Other linkage methods are: *centroid linkage* (using the average value of all objects in a cluster as the reference point for distances to other objects or clusters), *average linkage* (averaging all distances between pairs of objects in different clusters to decide how far apart they are) and *median linkage* (which is similar to average linkage except it uses medians).

In order to produce partitioned clusters K-MEANS clustering is the mostly used method.[23] It searches for the best way to divide objects into different sections so that they are separated as well as possible. The number of clusters to be produced has to be set in advance. The program then starts by selecting 'seed' objects for each cluster, which are spread as much as possible from the centre of all cases. Then it assigns all objects to the nearest seed. Next, reassignment of the objects takes place in order to reduce the within-groups sums of squares.

An example of cluster analysis on Fourier coefficients of IR spectra of polymers is described by Zupan.[24] Frankel[25] applied clustering to the library of IR spectra obtained from the USEPA.

2.2.2 Kohonen Networks. Kohonen networks[26] belong to the category of unsupervised classification techniques. Hence, they only require a set of input patterns. In

[23] J. A. Hartigan, 'Clustering Algorithms', John Wiley and Sons, New York, 1975.
[24] J. Zupan, *Anal. Chim. Acta*, 1982, **139**, 143.
[25] D. S. Frankel, *Anal. Chem.*, 1984, **56**, 1011.
[26] T. Kohonen, *Biol. Cybern.*, 1982, **43**, 59.

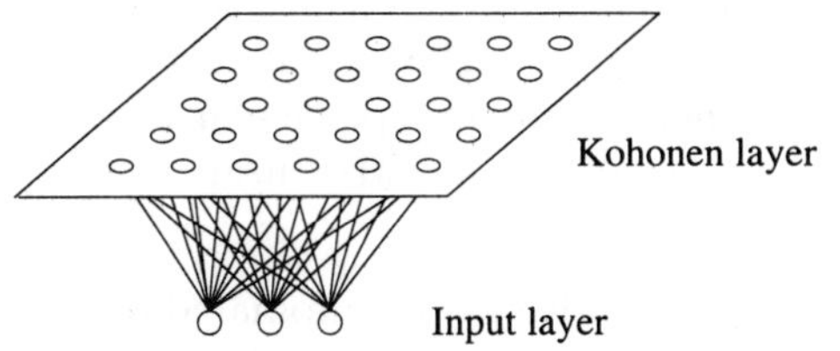

Figure 7 *Graphical representation of a Kohonen network*

the learning phase the network self-organizes the input data, thus reflecting the structure of input space. A Kohonen network consists of two layers, an input and an output layer. Each unit in the output layer is connected to all input units and has a corresponding weight vector. The units within the layers are not connected. As the input layer merely functions as a flow-through layer for the data it is often left out of graphical representations. The units in the output layer are usually ordered in a low-dimensional framework (commonly one- or two-dimensional). An example is given in Figure 7.

During training for each output unit the Euclidean distance of the corresponding weight vector and the input vector is calculated. The unit with the lowest value wins this competition. Next, the neighbourhood of the winning unit is determined. This consists of those units that are close to the winning unit in the output layer grid, *i.e.* within a square of a pre-defined size around the winning unit. Weights are updated for all units that are within this square according to:

$$\Delta w_{ij} = \eta \cdot (o_i - w_{ij}),$$

where w_{ij} is the weight of input unit i to output unit j and o_i is the input from unit i. The adjustment results in the winning unit and its neighbourhood having their weights modified and becoming more like the input pattern. Hence, when a similar pattern is presented, the unit will more likely become the winner.

Two parameters must be specified: the learning rate η and the size of the neighbourhood. The initial value η is usually set between 0.2–0.5 and is decreased during training. The neighbourhood width is initially set to a large value and is also decreased during the training phase.

An example of the application of a Kohonen network to the clustering of IR spectra has been described by Melssen *et al.*[27]

2.3 Spectrum Representation

The representation of the data is crucial to the appearance of a spectrum in pattern space. Not only the amount of data, but also the scales determine the relative

[27] W. J. Melssen, J. R. M. Smits, G. H. Rolf and G. Kateman, *Chemom. Intell. Lab. Syst.*, 1993, **18**, 195.

positions of patterns and thus the presence and appearance of clusters. When dealing with spectral data the inputs can be transmittance or absorbance values measured at distinct wavenumbers. The output can be any parameter related to the spectrum, *i.e.* the concentration of a particular component, the activity of a drug or, when used as a classifier, the structural class to which the respective compound belongs.

2.3.1 Coding Schemes. In the early days of automated spectrum interpretation use was made of very compact coding schemes for the representation of spectra. The first application of computerized classification methods for the interpretation of IR spectra was described by Kowalski *et al.*[2] They used spectra recorded in 0.1 μ bands from 2.0–14.9 μ giving a total of 130 pattern dimensions. The data were scaled on the strongest band which was assigned an intensity of 3.0. The largest peak in a 1.0 μ band was set to 2.0 and the remaining peaks were set to 1.0. No peak in a given 0.1 μ band resulted in a value of 0.0. Liddell and Jurs[28] also divided the spectra into intervals of 0.1 μ. The transmittance in each interval was converted into absorbance and put onto a 0–9 scale. Comerford *et al.*[29] divided the spectrum into 112 bands with widths of 50 cm^{-1} for the region 4000–2000 cm^{-1} and 25 cm^{-1} for the region 2000–200 cm^{-1}. Intensities were transformed from absorbance units to a scale from 0–9 depending on the intensity relative to the strongest band. They also applied linear discriminant functions to both Raman and IR spectra. As expected they found optimal classification results for concatenated IR and Raman data.

As computers became more powerful and FTIR spectrometers allowed easy capture of digitized spectral data, more detailed representations were used. Many applications of multivariate methods applied to full spectra with datapoint resolutions varying from 4–256 cm^{-1} and intensities with a precision of at least three decimals have been reported.

Fessenden and Györgyi[30] investigated the effect of using absorbance instead of transmittance data and of changing from wavenumbers to microns. They found that the wavelength scale does not significantly affect the results in contrast to the intensity scale. Transmittance values are to be preferred over absorbance data. They also investigated the use of different numbers of data points to represent the spectra ranging from a data point resolution of 8–256 cm^{-1}. Results started to deteriorate significantly when the data point resolution was chosen broader than 66 cm^{-1}.

2.3.2 Feature Selection. Feature selection may improve the results by discarding all input features (*i.e.* wavenumbers) that are irrelevant for the separation of the classes. Several methods have been applied to determine which features to remove and which to retain. Delaney *et al.*[31] apply feature reduction during the calculation of discriminating surfaces. They eliminate all weights that have different signs in the final weight vector when comparing the two training sessions with all weight

[28] R. W. Liddell III and P. C. Jurs, *Appl. Spectrosc.*, 1973, **27**, 371.
[29] J. M. Comerford, P. G. Anderson, W. H. Snyder and H. S. Kimmel, *Spectrochim. Acta*, 1977, **33A**, 651.
[30] R. J. Fessenden and L. Györgyi, *J. Chem. Soc., Perkin Trans. 2*, 1991, 1755.
[31] M. F. Delaney, P. C. Denzer, R. M. Barnes and P. C. Uden, *Anal. Lett.*, 1979, **12**, 963.

components set to a positve and all to a negative value. Alternatively, the previously mentioned weight vector obtained from the average class and non-class spectra is used. All weights with an absolute value less than a threshold (*e.g.* 0.1) are discarded. The threshold is increased and the process repeated until the classes are no longer separable.

Preuss and Jurs[32] applied feature reduction by subsequently leaving out one of the pattern descriptors. By observing the change in the discriminant vector, the importance of the feature is established. Thus a significant improvement in computing efficiency could be obtained without loss of predictive ability. A plot of the relevant features gave insight into the spectral regions of importance for a particular functionality.

Liddell and Jurs[28] developed a method called MAX for developing binary pattern classifiers. After setting the components of the weight vector to 0.1, the positive member of the training set with the lowest dot product (p) and the negative member with the highest dot product (n) are determined. Next, all ratios of the features of p and n are calculated. All ratios greater than 1.0 are put in decreasing order into a feature set favourable to p, and all ratios less than 1.0 in a set favourable to n. Ratios equal to one are discarded. These two lists are used to alter the weight vector. The top feature from each list is taken and the corresponding weight component is altered by $+1$ for the positive member and -1 for the negative one. Depending on the classification properties of the new weight vector this process is repeated with new features from the two lists.

Domokos *et al.*[7] applied several feature reduction methods. Important features were selected on the basis of their Fisher weight (*i.e.* the square of the difference between category means divided by the sum of the category variances), their variance weight (*i.e.* the ratio of inter- and intra-class variances), and the value of the weight vector components with respect to a pre-defined threshold. It appeared that the last method gives better results than the first two, although differences are not very large.

Heite *et al.*[33] describe the use of information theory to select the most discriminating wavelengths.

2.3.3 Principal Component Analysis. In order to reduce the number of data points to be used in the analyses without losing relevant information, principal component analysis (PCA) can be applied to the data (*i.e.* an eigenanalysis of the covariance matrix). Visual investigations using PCA as a pre-processing step were performed by Frankel.[25]

Perkins *et al*,[34–37] investigated the use of PCA for classification of IR spectra. Data were auto-scaled and feature weighted before PCA was applied, thus amplifying the variables relevant to the classification and increasing their contribution to the first principal components. The feature weight is the ratio of

[32] D.R. Preuss and P.C. Jurs, *Anal. Chem.*, 1974, **46**, 520.
[33] F.H. Heite, P.F. Dupuis, H.A. van't Klooster and A. Dijkstra, *Anal. Chim. Acta*, 1978, **103**, 313.
[34] J.H. Perkins, E.J. Hasenoehrl and P.R. Griffiths, *Anal. Chem.*, 1991, **63**, 1738.
[35] E.J. Hasenoehrl, J.H. Perkins and P.R. Griffiths, *Anal. Chem.*, 1992, **64**, 656.
[36] J.H. Perkins, E.J. Hasenoehrl and P.R. Griffiths, *Chemom. Intell. Lab. Syst.*, 1992, **15**, 75.
[37] E.J. Hasenoehrl and P.R. Griffiths, *Appl. Spectrosc.*, 1993, **47**, 643.

the interclass variance to the sum of the intraclass variances. Each measurement is multiplied with its feature weight. Instead of using the calculated feature weights w the squared weights w^2 can be applied, thus emphasising the larger weights. Another modification is to subtract one from the weights. As the minimum feature weight that can be calculated for a variable equals one, subtraction of one removes that variable essentially from the data set. Perkins *et al.* found the weighting function (w^2-1) to be superior to w. They used a discriminant function of the type:

$$\mathrm{d_i}(\boldsymbol{x}) = \ln(|\Sigma_\mathrm{i}|) + (\boldsymbol{x}-\boldsymbol{\mu}_\mathrm{i})' \, \Sigma_\mathrm{i}^{-1} \, (\boldsymbol{x}-\boldsymbol{\mu}_\mathrm{i}) - 2 \ln P_\mathrm{i},$$

where $\boldsymbol{x}$ is the score representation of a spectrum in two-dimensions, $\boldsymbol{\mu}_\mathrm{i}$ is the mean vector for class i, Σ_i is the covariance matrix of the class i training set and P_i is the *a priori* probability of a sample being in class i. Alternatively, use was made of a sigmoid function as discrimination function. The parameters of this function (temperature, bias) were optimized using a simplex method.

2.3.4 Interferograms. Recently, Small *et al.*[38–40] investigated the use of interferograms for the detection and identification of atmospheric pollutants. By applying compound-specific (*e.g.* SF_6) digital filters, modified interferogram segments were obtained and subsequently classified by a discriminant function. This function was trained using 3000 interferograms and a fixed-increment perceptron algorithm. Weights were updated according to:

$$w_{\mathrm{i}+1} = w_\mathrm{i} + c\,[\Sigma \boldsymbol{x}_{\mathrm{p,k}} - \Sigma \boldsymbol{x}_{\mathrm{a,k}}],$$

where $\boldsymbol{w}_{\mathrm{i}+1}$ is the updated weight, $\boldsymbol{w}_\mathrm{i}$ is the current weight and c is a user-specified constant. The summations were calculated over those vectors $\boldsymbol{x}_p$ (SF_6 present) and $\boldsymbol{x}_a$ (SF_6 absent) that were misclassified by w_i. An automated procedure was used to adapt c after each iteration. If the current weight vector produced fewer misclassifications than the previous one, c was increased as:

$$c_{\mathrm{i}+1} = c_\mathrm{i}/[1-(m_\mathrm{p}+m_\mathrm{a})/n],$$

where m_p and m_a are the number of misclassified patterns and n is the total number of patterns in the training set. If, however, the current weight vector produced more misclassifications, the value of c was decreased as:

$$c_{\mathrm{i}+1} = c_\mathrm{i}[1-(m_\mathrm{p}+m_\mathrm{a})/n].$$

For a test set containing 1726 interferograms 95.7% were classified correctly, whereas the number of false positives and negatives was 3.4 and 4.4%, respectively.

[38] G. W. Small, S. E. Carpenter, T. F. Kaltenbach and R. T. Kroutil, *Anal. Chim. Acta*, 1991, **246**, 85.
[39] G. W. Small and S. E. Carpenter, *Appl. Spectrosc.*, 1992, **46**, 49.
[40] S. E. Carpenter and G. W. Small, *Appl. Spectrosc.*, 1992, **46**, 1790.

3 FUTURE TRENDS

The automation of spectrum interpretation will benefit from a continuing interest. It is expected that non-linear alternatives to neural networks will be the subject of investigations and that their usefulness for spectrum classification will be assessed. However, classification solely is not sufficient for a full interpretation of complex spectra. Clearly, digitized storage of advanced strategies and ways of reasoning as used by expert spectrometrists will be required to tackle the most interesting but also most difficult problems. Eventually, hybrid systems will appear containing both rule-based parts for storage of expert reasoning and multivariate classification modules for dedicated tasks such as recognition of global spectral profiles, local spectral patterns and characteristic band shapes. The combination of these approaches may ultimately lead to a system capable of approximating expert performance.

CHAPTER 7

Computer Methods in Vapour Phase Infrared Spectroscopy

B.J. VAN DER VEKEN

1 INTRODUCTION

Vapour phase infrared spectroscopy is often associated with the detailed unravelling of rotational fine structure superposed on vibrational transitions. For molecules that allow such an analysis, a wealth of information on the compound under study is obtained. Unfortunately, this high resolution analysis in practice is limited to fairly small molecules. The reason for this is that with increasing mass of the molecule, the vapour phase infrared spectra become increasingly more complex. First, only rarely do many-atom molecules show sufficient symmetry to simplify rotational behaviour. This means that one has to deal with asymmetric top molecules, which are the most general, but also the most complex type of rotors. Because the separation between successive rotational levels is inversely related to the moments of inertia, as the molecules get bigger, the rotational levels tend to crowd. A first consequence is that the rotational transitions will show smaller frequency differences. Secondly, when experiments are performed at, or near, room temperature, for heavy molecules far more rotational levels will be thermally populated, so that in a decreasing frequency interval, an increasing number of transitions will occur. Moreover, as the molecules get bigger, they tend to have more and more low-frequency vibrations such as deformations and torsions. Again, when the spectrum of such a compound is investigated at ordinary temperatures, low-lying vibrational levels are substantially thermally populated, and each of the differently vibrationally excited molecules will give rise to vibrational transitions on which its own slightly different rotational fine structure is superposed. Therefore, for molecules of some size, even if under high resolution the fine structure can be sufficiently resolved, the sheer complexity of the spectrum prevents a complete analysis. By their own nature, there are far more molecules of this kind that can be investigated in the vapour phase than there are molecules that allow a high resolution approach. Therefore it would appear that, unless sophisticated experiments such as jet-expansions are used to reduce dramatically the complexity of the observed spectra, for the majority of molecules the information embedded in the vibrational–rotational spectra has to remain inaccessible. This, luckily, is not quite the case. The molecules envisaged here, when investigated at low resolution, say 0.5 cm^{-1}, are well known to give rise to highly characteristic P, Q, R rotational structures. The qualitative appearance of these structures has been used for a long time, for instance, to symmetry-classify the vibrational transitions on which they are observed. It is the aim of this contribution to show that far more than this qualitative information can be deduced from low resolution contours. This is brought about, as will be shown below, by relatively simple computer simulations of the observed spectra.

2 THE MODEL

For most molecules which we want to investigate, little or no structural information is available, so the model used for the simulations must be as simple as possible. Therefore, the rigid rotor model is used. In this model, apart from assuming the Born–Oppenheimer approximation for the separation of the electronic from the nuclear motions, also the vibrations and rotations are regarded as separable. This means that vibration–rotation interactions such as centrifugal distortion and Coriolis coupling are neglected. The molecule in the vibrational ground state, and the molecule in the excited vibrational state are thus treated as rigid bodies, for which the rotational behaviour is described by the rigid rotor theory. Very often, the rotational characteristics, *i.e.* the rotational constants, of the two vibrational states involved will be taken to be identical. We call this approximation the harmonic rigid rotor simulation. In a number of cases, it will be seen useful to improve the quality of the simulations by allowing the upper vibrational state rotational constants to differ from those of the lower vibrational state. The models we use to calculate such differences are based on an anharmonic potential for the molecular vibrations, and, therefore, simulations at that level will be termed anharmonic rigid rotor simulations.

3 THE CALCULATION OF THE ENERGY LEVELS

In the field free space, the rotation of a body is not subject to potential energy, and the total energy is equal to the kinetic energy of rotation. For the rotation of a rigid body around some axis through its centre of gravity the kinetic and thus the total energy is given by:

$$E=\frac{1}{2}I\,\vec{\omega}^2 \tag{1}$$

In this expression I is the moment of inertia of the body for that particular axis of rotation, and ω is the angular velocity. The angular momentum of the body caused by this rotation is:

$$\vec{J}=I\,\vec{\omega} \tag{2}$$

The expression for the energy can be rewritten as:

$$E = \frac{\vec{J}^2}{2I} \tag{3}$$

Eqn. (3) describes the rotation of the molecule around one particular axis, and the value of I is valid for this axis only. Therefore, eqn. (3) is not very general. It can be generalized by writing the rotation as a sum of three other rotations, one around each axis of a cartesian frame. If this frame is arbitrarily chosen, the expression for the energy of rotation will contain all elements of the inertial tensor, *i.e.* it will contain as constants I_{xx}, I_{yy} and I_{zz}, but also I_{xy}, I_{xz} and I_{yz}. It can be shown,[1] however, that for each rigid body there exists one cartesian frame in which the cross terms I_{xy}, I_{xz} and I_{yz} disappear. This frame is called the principal axis system of the body. The axes of this system are assigned the symbols *a*, *b* and *c*. Using this frame, the energy of rotation becomes:

$$E = \frac{\vec{J}_\mathrm{a}^2}{2I_\mathrm{a}} + \frac{\vec{J}_\mathrm{b}^2}{2I_\mathrm{b}} + \frac{\vec{J}_\mathrm{c}^2}{2I_\mathrm{c}} \tag{4}$$

In this expression, $\vec{J}_\mathrm{i}$ is the angular momentum around the axis *i* of the principal axis system, and I_i is the moment of inertia of the body around the axis *i*. The I_i are called the principal moments of inertia. For their calculation, a molecule is treated as an assembly of point masses, in which each nucleus is assigned the total mass of the atom it belongs to. Thus, the contributions owing to the electron clouds are neglected. The principal moment around axis *i* of an *n*-atomic molecule therefore is given as:

$$I_i = \sum_{\alpha=1}^{n} m_\alpha \left(r_\alpha^i \right)^2 \tag{5}$$

where m_α is the mass of atom a, and r_α^i is the distance of this atom to the axis *i*. The SI unit for this quantity is kg m²,[2] but it is a spectroscopic habit to use relative atomic masses for m_α, and to express r_α^i in Ångströms. The units of I_i then are amu × Å². By convention, the labels a, b and c are assigned to the principal axis in such a way that:

$$I_\mathrm{a} \leq I_\mathrm{b} \leq I_\mathrm{c}$$

In rotational spectroscopy, instead of the principal moments of inertia, usually the rotational constants A, B and C are used, which are defined as:

$$\mathrm{A} = \frac{h}{8\pi^2 c I_\mathrm{a}};\ \mathrm{B} = \frac{h}{8\pi^2 c I_\mathrm{b}};\ \mathrm{C} = \frac{h}{8\pi^2 c I_\mathrm{c}} \tag{6}$$

The dimension for these quantities is (length)$^{-1}$. Therefore, by expressing *h*, *c* and

[1] P. W. Atkins, 'Physical Chemistry', Oxford University Press, Oxford, 3rd edn., 1986.

[2] I. Mills, T. Cvitaš, K. Homann, N. Kallay and K. Kuchitsu, 'Quantities, Units and Symbols in Physical Chemistry', Blackwell Scientific Publications, 1988.

I_a in the appropriate derived SI units, the rotational constants can be given in the standard IR frequency unit, the wavenumber, cm^{-1}. Using the rotational constants, the energy of rotation for a general rotor is given as:

$$\frac{h}{4\pi^2 c} E = A\vec{J}_a^{\,2} + B\vec{J}_b^{\,2} + C\vec{J}_c^{\,2} \tag{7}$$

Because of the convention about the expression I_i, the rotational constants show the property:

$$A \geq B \geq C$$

It is well known that several classes of rotors can be distinguished, depending on the relative values of A, B and C. For this chapter we are interested in two classes only, *i.e.* the symmetric and the asymmetric top. For the former, two rotational constants are equal. Depending on the case such rotors are called prolate ($A > B = C$) or oblate ($A = B > C$). For the asymmetric top the three rotational constants differ from each other. The degree of asymmetry is usually expressed by Ray's asymmetry parameter κ:[3]

$$\kappa = \frac{2B - A - C}{A - C} \tag{8}$$

It is easily shown that when B approaches, C, κ will tend to a limiting value -1, which is reached for a prolate symmetric top. On the other hand, when B approaches A, κ tends to its other limiting value, $+1$. The latter is reached for an oblate symmetric top. Therefore asymmetric tops with a value of κ between -1 and 0 are called prolate asymmetric tops, while a rotor for which κ falls between 0 and $+1$ is called an oblate asymmetric top.

By the rules of quantum mechanics, the expression for the total energy of rotation is transformed into the Hamilton operator $\hat{H}$ by substituting the angular momenta J_i with the appropriate operators for angular momentum, for which we will use the symbols $\hat{J}_i$. When we integrate the constant $h/4\pi^2 c$ into $\hat{H}$, the following Hamilton operator results from eqn. (7):

$$\hat{H} = A\,\hat{J}_a^2 + B\,\hat{J}_b^2 + C\,\hat{J}_c^2 \tag{9}$$

This Hamiltonian is deceptively simple. It indeed turns out that the Schrödinger equation for this Hamiltonian:

$$\hat{H}\psi = E\psi \tag{10}$$

cannot be solved analytically for the general case in which A, B and C differ from each other. Before we discuss the way in which the above Schrödinger equation is solved, it is useful to observe that the Hamiltonian represents an angular

[3] B. S. Ray, *Z. Physik*, 1932, **78**, 74.

momentum problem. Therefore we will shortly review some of the characteristics of the latter.

The problem of angular momentum is treated in quantum mechanics in a very general way.[4] In particular, it is shown that the operator $\hat{\mathcal{J}}$ for the total angular momentum, the operator $\hat{\mathcal{J}}_c$ for one component of the angular momentum in a molecular axis system (for which we have quite arbitrarily chosen the *c* component in the principal axis system), and the operator $\hat{\mathcal{J}}_z$ for one component of the angular momentum in the laboratory axis system (for which we have again quite arbitrarily chosen the *z* component) commute among each other. Therefore, these three operators must have simultaneous eigenfunctions. The latter turns out to be a function of three quantum numbers which are called J, K and M. Therefore, the eigenfunctions are noted as |J, K, M>. Consequently it is shown that for the above operators the following eigenvalue equations hold:

$$\hat{\mathcal{J}}^2 \,|J, K, M\rangle = J(J+1)\,|J, K, M\rangle \tag{11}$$
$$\hat{\mathcal{J}}_c \,|J, K, M\rangle = K\,|J, K, M\rangle \tag{12}$$
$$\hat{\mathcal{J}}_z \,|J, K, M\rangle = M\,|J, K, M\rangle \tag{13}$$

This means that the eigenvalue of the total angular momentum, in units of $\hbar$, is $\sqrt{J(J+1}$, while eigenvalues of the chosen component in the principal axis, and of the chosen component in the laboratory frame, in units of $\hbar$, are K and M, respectively.

Unfortunately, the operators for the other two components of $\vec{\mathcal{J}}$ in the principal axes do not commute with $\hat{\mathcal{J}}$, and therefore the eigenfunctions |J, K, M > of $\hat{\mathcal{J}}$ are not eigenfunctions of, in this case, $\hat{\mathcal{J}}_a$ or $\hat{\mathcal{J}}_b$. It is shown[4] that the action of these operators on |J, K, M> is described by:

$$(\hat{\mathcal{J}}_a \pm \hat{\mathcal{J}}_b)\,|J, K, M\rangle = \sqrt{J \pm K)(J \pm K+1)}\,|J, K\pm 1, M\rangle \tag{14}$$

In view of what follows, it must be emphasized that in the above we have followed the generally adapted practice of identifying $\hat{\mathcal{J}}_c$ with the operator that commutes with J, but that we may as well identify either $\hat{\mathcal{J}}_a$ or $\hat{\mathcal{J}}_b$ with this operator (keeping in mind that only one component commutes with $\hat{\mathcal{J}}$, the other two not).

Also the operators for the other two components of $\vec{\mathcal{J}}$ in the laboratory axis fail to commute with $\hat{\mathcal{J}}$ and a result similar to eqn. (14) is obtained for the action $(\hat{\mathcal{J}}_x \pm \hat{\mathcal{J}}_y)$ on |J, K, M>.

Let us now investigate how these properties are used to solve the quantum mechanical problem of a rigidly rotating molecule. To this end, we rewrite the Hamiltonian eqn. (7), using the identity : $ax + by \equiv \frac{1}{2}(a+b)(x+y) + \frac{1}{2}(a-b)(x-y)$:

$$\hat{H} = A\hat{\mathcal{J}}_a^2 + \frac{1}{2}(B+C)(\hat{\mathcal{J}}_b^2 + \hat{\mathcal{J}}_c^2) + \frac{1}{2}(B-C)(\hat{\mathcal{J}}_b{}^2 + \hat{\mathcal{J}}_c^2) \tag{15}$$

[4] R. Zare, 'Angular Momentum, Understanding Spatial Aspects in Chemistry and Physics', J. Wiley and Sons, New York, 1988.

This expression can further be modified as follows:

$$\hat{H}=\frac{1}{2}(\mathrm{B}+\mathrm{C})(\hat{\mathcal{J}}_a^2+\hat{\mathcal{J}}_b^2+\hat{\mathcal{J}}_c^2)-\frac{1}{2}(\mathrm{B}+\mathrm{C})(\hat{\mathcal{J}}_a^2+\mathrm{A}\hat{\mathcal{J}}_a^2+\frac{1}{2}(\mathrm{B}-\mathrm{C})(\hat{\mathcal{J}}_b^2+\hat{\mathcal{J}}_c^2) \tag{16}$$

The sum of the squares of the components of the angular momentum in the principal axis system is equal to the square of the total angular momentum, and the same must be true for the corresponding operators. Therefore, the above expression can be written as:

$$\hat{H}=\frac{1}{2}(\mathrm{B}+\mathrm{C})\hat{\mathcal{J}}^2+\left[\mathrm{A}-\frac{1}{2}(\mathrm{B}+\mathrm{C})\right]\hat{\mathcal{J}}_a^2+\frac{1}{2}(\mathrm{B}-\mathrm{C})(\hat{\mathcal{J}}_b^2-\hat{\mathcal{J}}_c^2) \tag{17}$$

We first consider the special case of a prolate symmetric top. For this type of rotor, B equals C, and the above general Hamiltonian reduces to:

$$\hat{H}\ (\text{prol. sym. top})=\mathrm{B}\hat{\mathcal{J}}^2+(\mathrm{A}-\mathrm{B})\hat{\mathcal{J}}_a^2 \tag{18}$$

This Hamiltonian contains the operator $\hat{\mathcal{J}}$ and the operator for one component of the angular momentum, and we can choose this component to commute with $\hat{\mathcal{J}}$. Because no other components of $\hat{\mathcal{J}}$ are present in the Hamiltonian for the prolate symmetric top, it follows that the functions |J, K, M> are also the eigenfunctions for the prolate symmetric top. Analogously, by grouping terms in the general Hamiltonian eqn. (7), it is easily shown that the |J, K, M> are eigenfunctions for the oblate symmetric top as well. Thus, in general, the functions |J, K, M> are eigenfunctions for the symmetric top.

Using the properties from eqns. (11) and (12), the eigenvalues for the prolate symmetric top can be obtained:

$$\hat{H}\ (\text{prol. sym. top})\ |\mathrm{J},\mathrm{K},\mathrm{M}\rangle=\left[\mathrm{B}\hat{\mathcal{J}}^2+(\mathrm{A}-\mathrm{B})\hat{\mathcal{J}}_a^2\right]|\mathrm{J},\mathrm{K},\mathrm{M}\rangle \tag{19}$$

It follows for the energies:

$$E\ (\text{prol. sym. top})=\mathrm{BJ}(\mathrm{J}+1)+(\mathrm{A}-\mathrm{B})\mathrm{K} \tag{20}$$

and similarly for an oblate symmetric top:

$$E\ (\text{obl. sym. top})=\mathrm{BJ}(\mathrm{J}+1)+(\mathrm{C}-\mathrm{B})\mathrm{K} \tag{21}$$

For an asymmetric top, the last term in the Hamiltonian eqn. (16) does not disappear. Consequently, the Hamiltonian for this type of rotor contains the operators for all three components of $\vec{\mathcal{J}}$ in the principal axis system. It follows that the functions |J, K, M> are not eigenfunctions of the asymmetric top Hamiltonian. Also, in view of the properties in eqn. (14), it is clear that a closed expression for the energy levels similar to eqns. (20) and (21) cannot be obtained for the asymmetric top.

Because of the impossibility of an analytical solution, the quantum mechanical solution for the asymmetric top is solved using matrix mechanics. It will be appreciated that by neglecting the third term in eqn. (17), a Hamiltonian results that is formally identical to the one for a symmetric top. This suggests that a Heisenberg matrix $<\psi_i|\hat{H}|\psi_j>$ be set up in the symmetric rotor functions $|J, K, M>$. By using the properties from eqns. (11)–(14), and by writing $(\hat{\mathcal{J}}_b^2-\hat{\mathcal{J}}_c^2)$ as $(\hat{\mathcal{J}}_b+\hat{\mathcal{J}}_c)(\hat{\mathcal{J}}_b-\hat{\mathcal{J}}_c)$, the elements of the Heisenberg matrix can be calculated without problems. It can be seen from eqn. (14) that the operator $(\hat{\mathcal{J}}_b^2-\hat{\mathcal{J}}_c^2)$ does not couple functions belonging to different values of the rotational quantum number J. Therefore, the Heisenberg matrix block-diagonalizes, and only finite matrices, one for each value of J, must be diagonalized to obtain the energy levels for the asymmetric top. Because of the properties of K, these matrices are of the dimension $2J+1$. As usual, the eigenvectors of the Heisenberg matrix are the eigenfunctions of the rotor.

The matrix described above is not used in practical calculations, because the problem can be simplified by invoking symmetry. The Hamiltonian, eqn. (17), is invariant under twofold rotations about the *a*, *b* and *c* axes. Hence, the eigenfunctions of the asymmetric top must transform according to the irreducible representations of the point group D_2. The eigenfunctions $|J, K, M>$ of the symmetric rotor, however, show a different symmetry,[5] and, therefore, the Heisenberg matrix for a specific value of J does not block-diagonalize. Fortunately, it is possible to construct linear combinations of the $|J, K, M>$ basis that have the required symmetry. These combinations are formed by the Wang transformation[6] which is defined as:

$$\text{for } K\neq 0: |J, K, M> = \frac{1}{\sqrt{2}}\left[|J, K, M> + (-1)^s |H, -K, M>\right] \quad (22a)$$

$$\text{for } K=0: |J, 0, M, s> = |J, 0, M> \quad (22b)$$

In this expression, $s=0$ or 1.[6] When the Wang transformed basis is used, the elements of the Heisenberg matrix must be transformed as well. The exact expressions in the Wang basis are not essential for the present argument, and can be found in the literature.[6] What is essential is that when using the Wang basis, the Heisenberg matrix for each J block-diagonalizes into four smaller blocks, each of approximate dimension J/2 (compared to $2J+1$ for the original matrix). These subblocks are described as O^+, O^-, E^+ and E^-, and all the eigenstates belonging to the same block transform according to the same irreducible representation of D_2. With the diagonalization of a matrix requiring a computer central processing unit (CPU) time roughly proportional to the third power of the dimension of the matrix, the reduction of the Heisenberg matrix to four smaller blocks induces a substantial saving on CPU time (for large values of J, the reduction factor is approximately 16!). Moreover, the eigenvalues and eigenvectors are now automatically classified according to their symmetry, which is of advantage in the next step of the procedure, the calculation of transition intensities.

[5] H. C. Allen, Jr. and P. C. Cross, 'Molecular Vib-Rotors', J. Wiley and Sons, New York, 1963.
[6] S. C. Wang, *Phys. Rev.*, 1929, **34**, 243.

An important note should be made with respect to Coriolis coupling, which in the rigid rotor approximation used here is neglected. It is well known that for symmetric top molecules the components of degenerate modes show strong Coriolis interactions that greatly upset the rigid rotor energy level scheme. For asymmetric top molecules, because degeneracies caused by symmetry do not occur, strong Coriolis interactions are less likely. However, they are not excluded between modes with nearby frequencies. For such modes, rigid rotor calculations are of little value. Fortunately, the Coriolis coupling deforms the P and R branches, and disturbs their intensity ratio. Therefore, bands influenced by Coriolis coupling are recognized easily. Such bands have been avoided in the simulations that will be discussed in the second part of this chapter.

One final remark must be made on the practical calculations of energy levels; computer matrix diagonalization procedures are iterative, and converge faster as the non-diagonal elements are smaller. It follows from the properties of eqn. (14) that off-diagonal elements are generated by the third term of the Hamiltonian, eqn. (17). For prolate asymmetric tops, and certainly for prolate near-symmetric tops, the coefficient $(B-C)$ in these terms ensures that the off-diagonal element will be small. For oblate asymmetric tops in principle eqn. (9) can be rewritten such that the coefficient of the third term becomes $(A-B)$, which is small for this type of rotor. Because most of the asymmetric top molecules encountered in nature are prolate, and to avoid extra complications in the computer program that calculates the energy levels, the computer program is based on the prolate top Hamiltonian eqn. (17), at the cost of some CPU time when an oblate asymmetric top is investigated.

4 THE CALCULATION OF TRANSITION INTENSITIES

A band in a low resolution vapour phase infrared spectrum is a superposition of many vibration–rotation transitions, each of which takes place between a state n'' and a state n' of the molecule. Using time dependant perturbation theory, and when the wavelength of the photon is much larger than the size of the molecule, quantum mechanics arrives at the following expression for the intensity of an individual transition:[5]

$$I_{n''\rightarrow n'}=\frac{8\pi^3 \nu N g_{n''}[1\text{-}\exp(-h\nu/kT)]\exp(-E_{n''}/kT)}{3hc\left[\sum_n g_n\exp(-E_n/kT)\right]}\,|\vec{\mu}_{n'',n'}|^2 \tag{23}$$

For us, the essential part of this expression is $|\vec{\mu}_{n'',n'}|^2$, the matrix element of the dipole moment between the two states of the system. The factor that precedes this matrix element takes into account the radiation density of the environment of the molecule, and takes into account the thermal population of the initial state of the system. The matrix element $\vec{\mu}_{n'',n'}$ is defined as:

$$\vec{\mu}_{n'',n'}=\langle\psi(n'')|\vec{\mu}|\psi(n')\rangle \tag{24}$$

In this expression, $\psi(n'')$ and $\psi(n')$ are the molecular wave functions of the initial and final states, respectively. Within the approximations made, and because only rotational–vibrational transitions are studied, it is sufficient to treat $\psi(n'')$ and $\psi(n')$ as a product of a rotational and a vibrational wavefunction.

It is obvious that a molecular dipole moment changes with the orientation of the molecule, *i.e.* $\vec{\mu}$ is a function of the coordinates that describe the rotation of the molecule. Also, the dipole moment will change during the vibrations of the molecule, which means that $\vec{\mu}$ is a function of the normal coordinates as well. The dependance of $\vec{\mu}$ on the rotational and vibrational coordinates is treated in the following way. First, the dipole moment is written as a sum of components in the laboratory axis system:

$$\vec{\mu}=\sum_{F} \vec{\mu}_{\mathrm{F}} \qquad (F=x, y, z) \tag{25}$$

The dipole moment can also be described in terms of its components in the principal axes system:

$$\vec{\mu}=\sum_{g} \vec{\mu}_{\mathrm{g}} \qquad (g=a, b, c) \tag{26}$$

The components in the laboratory axis can then be described in terms of the components in the principal axes system by using the direction cosines Φ_{Fg} between the directions of the laboratory and principal axis systems:

$$\vec{\mu}_{\mathrm{F}}=\sum_{g} \Phi_{\mathrm{Fg}}\vec{\mu}_{\mathrm{g}} \tag{27}$$

and, therefore, we find for $\vec{\mu}$:

$$\vec{\mu}=\sum_{F,g} \Phi_{\mathrm{Fg}}\vec{\mu}_{\mathrm{g}} \tag{28}$$

As the principal axes rotate with the molecule, the components $\vec{\mu}_{\mathrm{g}}$ are independent of the coordinates used in the rotational factor of the molecular wave function, and are the only functions of the normal coordinates that are used in the vibrational factor of the molecular wave function. Therefore, by writing the molecular wave function as a product of a rotational and a vibration part:

$$\psi(n)=\psi_{\mathrm{r}}(n)\ \psi_{\mathrm{v}}(n)$$

The matrix element of the dipole moment can be written as:

$$<\psi(n'')|\vec{\mu}|\psi(n')=\sum_{\mathrm{F,g}} <\psi_{\mathrm{r}}(n'')|\Phi_{\mathrm{Fg}}|\psi_{\mathrm{r}}(n')><\psi_{\mathrm{v}}(n'')|\vec{\mu}_{\mathrm{g}}|\psi_{\mathrm{v}}(n')> \tag{29}$$

The dependence of $\vec{\mu}_g$ on the normal coordinates Q_i is expressed by developing $\vec{\mu}_g$ in a power series in the Q_i:

$$\vec{\mu}_g = \vec{\mu}_g^0 + \sum_{i=1}^{3n-6} \left(\frac{\partial \vec{\mu}_g}{\partial Q_i}\right)_0 Q_i + \ldots \tag{30}$$

The general experience is that the contributions of quadratic and higher order terms in the Q_i series are much smaller than those in linear terms, and are, therefore, neglected. Moreover, because of the orthogonality of wavefunctions, the linear part $\vec{\mu}_g^0$ cannot contribute the matrix element, eqn. (24). Finally, in this chapter we will almost exclusively study fundamental vibrational transitions, *i.e.* transitions in which only the vibrational quantum number of one normal vibration changes from 0 to 1. Owing to the properties of harmonic oscillator functions, from the summation over the normal coordinates all terms will cancel except the one that contains the normal vibration under consideration. Then, substitution of eqn. (30) into eqn. (29) leads to:

$$\vec{\mu}_{n'',n'} = \sum_{F,g} \left(\frac{\partial \vec{\mu}_g}{\partial Q_i}\right)_0 <\psi_r(n'')|\Phi_{Fg}|\psi_r(n')> <\psi_v(n'')|Q_i|\psi_v(n')> \tag{31}$$

The last factor expresses the vibrational selection rule. In the harmonic approximation made here, for the fundamental transition this factor is equal to 1. The other integral in expression (31) represents the rotational selection rule, and, as can be seen, this factor depends on *g*, one of the directions of the principal axes system. For instance, the value for this integral for a component $(\partial \vec{\mu}_a/\partial Q_i)_0$, which is oriented along the *a* axis, will be different from the value for the *b* or *c* component. This characteristic has an important consequence. When in a molecule, as a consequence of symmetry, the dipole gradients for the different normal modes are restricted to be parallel to one of the principal axes, the rotational contours show a distinct character, depending along which axis the gradient is oriented. These contours are called type A, B and C, for gradients parallel to the *a*, *b* and *c* axis, respectively. For reasons which will become clear in a following paragraph, in the simulation procedure transitions involving gradient components along different principal axes are calculated separately, and are collected separately to form the type A, B and C contours. These will be referred to as the pure type contours A, B and C, respectively.

Because the rotational wave functions transform according to the irreducible representations of D_2, it can be expected that symmetry considerations are important as to which transition will contribute to which of the pure type contours. This matter has been treated in detail in the literature,[5] and a full description is judged outside the scope of this chapter.

Once the intensities of the individual transitions have been calculated, they can be assembled into a pure type (A, B or C) spectrum. The frequency, in wavenumbers, at which a transition occurs is given by the energy difference of the final and initial states of the molecule. For the type of transitions under study, and for the approximations made, this difference is given as the sum of two terms. The first term is the energy difference between the rotational ground states of the final and initial vibrational states. This contribution is the vibrational energy, and it is the same for all the individual rotational transitions within one contour. Therefore, it is equated to zero. The second term is the difference between the energies of the final and the initial rotational levels, and is obtained as the difference between the corresponding eigenvalues of the rotational Hamiltonian, eqn. (17).

The contribution of an individual transition to a calculated spectrum can be described as a convolution of the intrinsic profile of the transition, with an instrument lineshape function. The intrinsic lineshape for an ordinary experiment is determined primarily by pressure broadening. This profile is usually taken to be Lorentzian. The full width at half height (FWHH) of the intrinsic lineshape will be much smaller than the instrument line shape, the FWHH of which is equal to the resolution at which a spectrum is investigated. Under these circumstances, the intrinsic lineshape has little or no influence on the result of the convolution. Therefore, the contribution of an individual transition is approximated by a band that has the instrument line shape, and of which the integrated intensity is calculated from expression (23).

When calculating spectra for comparison with experiments run on a Fourier transform (FT) spectrometer, the instrument lineshape function is determined by the type of apodization applied to the interferogram.[7] For comparison with experiments run on a dispersive instrument, the lineshape functions may be taken to be Gaussian.

Finally, the contribution of an individual transition must be added to the calculated spectrum. The latter is a vector, each element giving the intensity of the spectrum at a certain spectral frequency. These frequencies, for the ease of further manipulations, are chosen with a constant interval, for instance 0.25 cm^{-1}. The contribution of an individual transition to such a spectrum is obtained by positioning its chosen line shape, with adjusted intensity, at the transition frequency, and interpolating this line shape at the frequencies at which the calculated spectrum is determined. In theory, the instrument lineshape stretches from minus to plus infinity. However, its intensity rapidly decays with increasing distance from the band centre, and in practical calculations the interval in which an individual transition has a non-zero intensity is restricted to a small multiple (typically five) of its FWHH. The interpolated line shape intensities form a vector representing the individual transition, and this vector is added to the vector representing the calculated spectrum.

5 THE ORIENTATION OF THE DIPOLE GRADIENT

When the asymmetric top under study has sufficient symmetry, the dipole gradient $(\partial\vec{\mu}/\partial Q_i)_0$ for each of the normal modes is parallel to one of the principal axes. The transitions observed in the vapour phase infrared spectrum of such a compound will be pure type A, B or C transitions, that can be straightforwardly compared with the calculated type A, B or C transitions.

[7] R.J. Bell, 'Introductory Fourier Transform Spectroscopy, Academic Press, New York', 1972.

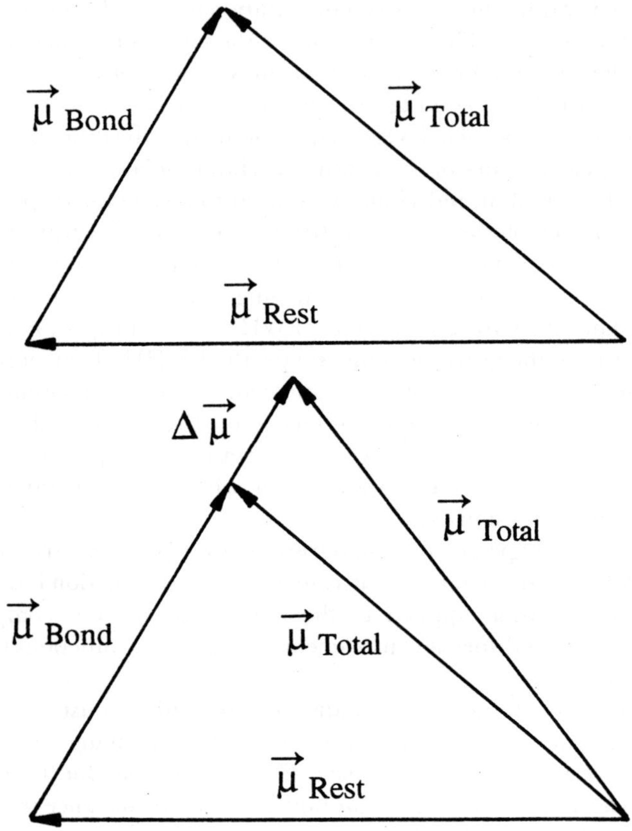

Figure 1 *The bond dipole model (BDM) for the orientation of the dipole gradient for a localized stretching mode*

In the general asymmetric top, however, the dipole gradients have arbitrary orientations with respect to the principal axes. The transitions observed in the spectra then are so-called hybrid bands. It can be shown[8] that a hybrid band is a linear combination of the pure type transitions of that rotor:

$$H(\bar{\nu}) = \cos^2\alpha\ \mathrm{A}(\bar{\nu}) + \cos^2\beta\ \mathrm{B}(\bar{\nu}) + \cos^2\gamma\ \mathrm{C}(\bar{\nu}) \tag{32}$$

In this expression, $H(\bar{\nu})$ is the hybrid contour, and $\mathrm{A}(\bar{\nu})$, $\mathrm{B}(\bar{\nu})$ and $\mathrm{C}(\bar{\nu})$ are the pure type transitions. The constants $\cos^2\alpha$, $\cos^2\beta$ and $\cos^2\gamma$ are the squares of the direction cosines of the dipole gradient in the principal axis system of the molecule. Thus, α, β and γ are the angle between $(\partial\vec{\mu}/\partial Q_i)_0$ for the vibration under consideration, and the *a*, *b* and *c* principal axes, respectively. From this it is clear that for the simulation of a hybrid band the orientation of the dipole gradient with respect to the principal axis system must be known. Unfortunately, there are no

[8] B. J. van der Veken, in 'Vibrational Spectra and Structure', ed. J. R. Durig, Elsevier, Amsterdam, 1986, vol. 15, p. 313.

simple rules from which this orientation can be deduced. As a consequence, at least at the outset, a model has to be used from which the orientation of the gradient can be predicted.

A very simple model is the bond dipole model (BDM), based on Thomson's hypothesis that a molecular dipole moment can be regarded to be the vector sum of bond dipole moments, *i.e.* of dipole moments associated with, and directed along, the individual chemical bonds of the molecule.

The use of the model is best illustrated by an example. Consider the molecule CHFClBr, of which we intend to simulate the C–H stretching fundamental. Because we are interested in the C–H bond, we assume that the dipole moment $\vec{\mu}_{total}$ of the molecule can be written as the vector sum of the bond dipole of the C–H bond, $\vec{\mu}_{bond}$, and a dipole moment $\vec{\mu}_{rest}$ which is the sum of all other bond moments of the molecule. This situation is shown in the top part of Figure 1. When the C–H bond is stretched, its bond dipole moment will change to a new value ($\vec{\mu}_{bond} + \Delta\vec{\mu}$), also oriented along the C–H bond. Therefore, $\Delta\vec{\mu}$ must be parallel to the C–H bond. It is further assumed that the stretching of the C–H bond does not influence the bond moments of the other bonds. If this is the case, $\vec{\mu}_{rest}$ is unaffected by the stretching of the C–H bond, and the resulting situation is shown in the lower half of Figure 1. Because $\vec{\mu}_{rest}$ is unaffected, $\Delta\vec{\mu}$ not only represents the change in the C–H bond dipole, but is also equal to the change of the total dipole moment. When taken for a unit of C–H stretching in the vicinity of the equilibrium bond length, the change of the total dipole moment is a good approximation to the dipole gradient, $(\partial\vec{\mu}/\partial Q_i)_0$. (It is not quite identical to it, however: during the stretching of the C–H bond, the molecule moves in an Eckart frame, and this in general involves a reorientation of the rest of the molecule relative to the principal axis frame. This in its turn reorients $\vec{\mu}_{rest}$ relative to the axis system, and this affects the orientation of $(\partial\vec{\mu}/\partial Q_i)_0$; this contribution is neglected in the BDM.) Hence, the BDM predicts that for a stretching localized in a single bond, the dipole gradient will be parallel to the bond. For most molecules, by comparing to similar molecules, or from *ab initio* calculations, a reasonable molecular structure can be constructed. Once this structure is known, the orientation of the chemical bonds in the principal axes can easily be calculated. It then follows that the BDM orientations of dipole gradients can be obtained from simple structural considerations.

The above model is easily expanded to include coupled vibrations such as symmetric or antisymmetric XY_n stretches: the gradient will be parallel to the XY_2 bisector for the symmetric stretch, and perpendicular, but in the XY_2 plane, for the antisymmetric stretch. For the XY_3 symmetric stretch, the gradient is predicted to be parallel to the threefold axis of the grouping, *etc.*

Of course, the BDM is subject to criticism. For instance, any normal vibration of a molecule must leave its centre of mass invariant. When stretching a bond, other atoms generally require rearrangement and this will affect $\vec{\mu}_{rest}$. However, it is known that some characteristic vibrations like those of C–H bonds are energetically highly localized, and therefore involve only minor adjustments to the rest of the molecule. In such cases the BDM predictions may be hoped to work well. Some applications, and the possible sources of its failure, of the BDM will be discussed in the second part of this chapter.

6 FACTORS CAUSING THE BROADENING OF SPECTRA

It is rather frustrating to compare a freshly calculated hybrid with an experimental vapour phase infrared contour: almost invariably the experimental spectrum will be substantially broader than the calculated one. This broadening is especially clear for the Q branch, which in the harmonic approximation is calculated to be very narrow (for the much broader P and R branches the broadening is somewhat more difficult to detect, simply because of their own width). There are several reasons for this broadening.

First, in a real molecule the rotational constants of the upper and lower vibrational states are not identical; these differences cause small shifts in individual transitions which in the harmonic approximation are degenerate. At low resolution, this is observed as a broadening of the Q branch.

The usually more important broadening, however, is due to the presence of excited molecules in the sample studied. In a polyatomic molecule, a number of deformational and torsional modes are present. These vibrations in general appear at low frequencies, and at room temperature a substantial fraction of a sample will consist of molecules that have been thermally excited into these low-lying modes. The frequency at which a stretching mode occurs in a thermally excited molecule is slightly shifted from its value in a molecule in which the initial state is the vibrational ground state. Thus, when a complex compound is investigated at normal temperatures, an observed band in its IR spectrum will be a superposition of a substantial number of bands, each originating in a different kind of thermally excited molecule. Sometimes these shifts are a couple of wavenumbers or more: in those cases the experimental Q branch will show a more or less resolved component for each of the excited states. A pronounced example of that is found in the C–H stretching region of CHD_2OCD_2F.[8] The effect is obvious only in the Q branch because the P and R branch are always much broader than the Q branch and, therefore, cannot show resolved components for shifts of a few wavenumbers. More often, however, the shifts between the bands in different thermally excited molecules are a mere few tenths of a wavenumber. Even in the Q branch the different components then no longer are resolved: the thermal structure in those cases only broadens the contour, which, again, is most obvious for the Q branch. It may be remarked that for thermal excitation in some deformational modes an upward shift of the vibrational transition under study results, while for thermal excitation in other modes, a downward shift is induced. As a result, broadening occurs on both sides of the band centre, and a broad but fairly symmetric Q branch results. Unfortunately, there exists no simple theory (or not even a complex one) from which the shifts occurring in excited states can be predicted. As a consequence, when the thermal structure of a band is unresolved, it can be accounted for in an approximate way only. In the first place it is assumed that the rotational constants of the thermally excited molecules are identical to those of the vibrational ground state, *i.e.* the harmonic approximation is adopted. Secondly, it is assumed that the dipole gradient for a transition in a thermally excited molecule is the same as for the transition occurring in a ground state molecule. Under these assumptions, the experimental spectrum is a superposition of identical contours, with adjusted relative intensities, slightly shifted relative to each other. Such a pattern can be described as a convolution of the ground state contour with another function (see Chapter 3), that describes the distribution of the excited state contours. For simplicity, for the latter usually a Gauss–Lorentz sum profile is used. The FWHH of this function is chosen such that the FWHH of the Q branch of the convolved theoretical band is equal to the experimental Q branch width. When the thermal structure is resolved, also the harmonic approximation is adopted, but now a complex hybrid can be constructed that contains a simple hybrid, with adjusted relative intensity, at each of the frequencies at which experimentally a resolved component is observed. Usually, not all thermal structure is resolved, so that a complex hybrid may still require a convolution before it reproduces the experimental widths of the Q branches.

7 COMPUTATIONAL DETAILS

A full simulation is performed using two separate computer programs. The first program calculates the pure type A, B and C spectra for the molecule under study. Input to this program are, in the first place, the rotational constants of the vibrational ground and excited states. Other information required by this programe is J_{max}, the maximum value of the rotational quantum number J for which the eigenvalues are calculated. From eqn. (17) it is clear that the energy of a rotational level belonging to a specific value of J, contains the contribution $\frac{1}{2}(B+C)J(J+1)$. Hence, the energy of a rotational level increases roughly as the square of J. A transition between a rotational level of the vibrational ground state and one of the excited vibrational state can contribute measurable intensity to a theoretical band only when the lower level has a measurable thermal population. Since the energies of the rotational levels depend on J^2, it follows that a calculation can be limited to J_{max}, for which the populations are sufficiently low. As the energy also depends on the factor $\frac{1}{2}$ (B+C), the choice of J_{max} depends on the molecule under study. For lighter molecules, for which B and C have large values, a J_{max} of 80–100 may be sufficient, but for bigger molecules converged contours are obtained only when using J_{max} as high as 200. It must be emphasized that the choice of J_{max} is not without importance, as the amount of CPU time of the first program, as has been said before, approximately increases to the third power of J_{max}, and for high values of the latter, this time becomes quite respectable. For instance, the calculation of the pure type A, B and C contours for a typical molecule to a J_{max} of 180 requires some 30 min of CPU time on an IBM Risc 6000/355.

Further input to the first program consists of the frequency interval around the band centre in which the pure type spectra are to be calculated, and the (constant) frequency interval between two successive spectral points. Finally, because of the Boltzmann factor determining the thermal population of the rotational levels of the vibrational ground state, the temperature at which the spectra are to be calculated must also be specified.

The second program used in a simulation is a calculational and graphical package. From the pure type A, B and C contours calculated by the first program, this package allows the construction of a single hybrid from the direction cosines of its dipole gradient, the construction of complex hybrids containing several excited states, and the construction of spectral regions containing different hybrids at different frequencies. This package also allows the convolution of hybrids to account for unresolved thermal structures, the adding or subtracting of a baseline, and the rescaling of spectra. Finally, this package has a number of graphical functions. The constructed hybrids can be inspected on screen or in hard copy, and this package also allows the superposition of experimental and theoretical spectra.

The success of a simulation is measured by the goodness of the fit between the experimental and simulated spectra. For the sake of argument, we assume that the experimental spectrum consists of a single P,Q,R structure, but the generalization to more complex cases will be obvious. The goodness of the fit is judged from a superposition of experimental and simulated contours. When the experimental contour has a central Q branch, the superposition is done in such a way that the Q branch maximum of the calculated band coincides with that of the experimental band. This requires rescaling of the calculated intensities. Because the simulation concerns the evolution of the contour as a function of frequency, while the integrated intensity of the contour is not the object of study, this rescaling is of no importance to the result. A typical characteristic of a P,Q,R contour is the ratio of the intensity of the Q branch to the intensity of the P or R branch. Comparing the calculated relative P or R intensities to the experimental ones is possible only when the integrated Q branch intensities of experiment and simulation are the same. Having equalized the band centre intensities in the superposition, the important factor determining the integrated intensity of the Q branch is its FWHH: by convolving the calculated contour it is attempted to reproduce the experimental Q branch FWHH. The profile of the experimental Q branch is determined by the spread of the excited states and, to some extent, is also affected by factors such as pressure broadening, and instrumental parameters such as the apodization used. To account for this, the convolution can be performed with Gauss–Lorentz sum profiles, the exact composition of which is determined by comparing the wings of calculated and experimental Q branches. Once the Q branch of the simulated contour matches as closely as possible its experimental counterpart, the goodness of the simulation can be evaluated. In this, three characteristics are used. First, as stated above, the relative intensities of the P or the R branches are compared: evidently, when the calculated relative intensities differ strongly from the experimental values, the simulation was unsuccessful, and the model used is flawed. A second criterium is the PR separation, which is the frequency difference between the maxima of P and R branches. This characteristic depends on the orientation of the dipole gradient (because type A, B and C have slightly different PR separations), but, more importantly, also depends, in an implicit way, on the structure of the molecule.[9] Therefore, the PR separation of the calculated hybrid in some cases is a powerful argument in the acceptance or rejection of a simulation. Thirdly, the intensity evolution as function of frequency of the P and R branches also is of importance. If the contour in the far wings of the calculated P and R branches is below the experimental one, the difference may be caused by the

[9] W. A. Seth Paul, *J. Mol. Struct.*, 1969, **3**, 403.

presence, in the experimental spectrum, of weak transitions such as combination bands or overtones in these regions. However, in case the calculated wings are significantly higher than the experimental ones (small differences may be due to neglecting of centrifugal distortion in the calculation), with some certainty the model used in the simulation can be rejected.

Finally, before proceeding to a number of examples, it must be remarked that a quantification of the goodness of a fit would be very useful. It is, from the above, clear that several criteria that measure rather different aspects of a contour are simultaneously involved in the evaluation. As a consequence, reliable, generally applicable quantitative measures are difficult to define. Because of this, the qualitative judgement of the goodness by an experienced spectroscopist usually is more reliable than a judgement based on the blind application of some quantitative rule. Exceptions,[10] as usual, are taken to confirm this point of view.

8 THE USE OF ROTATIONAL FINE STRUCTURE

When the rotational constants of the molecule under investigation are large enough, sometimes rotational fine structure is observed on some or all of the vapour phase infrared absorption bands of the molecule. Under favourable conditions this fine structure can be exploited in structural studies and, therefore, it is of interest to know the rules governing the frequency separation between successive fine structure maxima. As will be seen, regular rotational fine structure can only be observed when the rotor under study approximately is a symmetric top, *i.e.* when its κ value either is close to -1 (prolate top) or $+1$ (oblate top), and it is useful to review the characteristics of the symmetric top fine structure.

Symmetric tops give rise to parallel and perpendicular rotation–vibration contours in their vapour phase infrared spectra. When the dipole gradient $(\partial\vec{\mu}/\partial Q_i)_0$ for a normal coordinate Q_i is parallel to the unique principal axis of the top, the contour observed for the fundamental transition involving Q_i is called the parallel contour, while in the other case, when the gradient is perpendicular to the unique axis, the perpendicular contour is observed. The different contours, of course, are the consequence of different selection rules for the rotational transitions.[5] When the Hamiltonian for a symmetric rotor is symbolically written as:

$$\hat{H} = \alpha\hat{J}^2 + \beta\hat{J}_a^2 \tag{33}$$

then it is easy to show[5] that the frequency separation between two successive fine structure lines of the P or R branch of a parallel band, in the harmonic approximation equals 2α. This amounts to 2B both for the prolate and oblate cases. For a perpendicular band, the frequency difference between two successive Q branches of a perpendicular band, also in the harmonic approximation, is equal to 2β. For a prolate rotor 2β is equal to $2(A-B)$, for an oblate rotor it is equal to $2(C-B)$.

To see how for near-symmetric tops, *i.e.* asymmetric tops for which $|\kappa|$ is nearly

[10] B.J. van der Veken, W.A. Herrebout and H.H. Liefooghe, *J. Mol. Struct.*, 1992, **268**, 293.

1, the rotational fine structure behaves, we use as an example a prolate asymmetric top. For the latter the rotational Hamiltonian was given in eqn. (17):

$$\hat{H}=\frac{1}{2}(B+C)\hat{\mathcal{J}}^2+[A-\frac{1}{2}(B+C)]\hat{\mathcal{J}}_a^2+\frac{1}{2}(B-C)\left(\hat{\mathcal{J}}_b^2-\hat{\mathcal{J}}_c^2\right) \tag{17}$$

When this top is near-symmetric, B and C are nearly equal, and the third term of the above Hamiltonian becomes very small. When this term is neglected, the Hamiltonian reduces to the one for the prolate symmetric top, with the following values for α and β:

$$\alpha=\frac{1}{2}\ (B+C) \tag{34}$$

$$\beta=\left[A-\frac{1}{2}\ (B+C)\right] \tag{35}$$

Thus, for prolate near-symmetric tops rotational fine structure is expected that shows the characteristics of the limiting symmetric top. In a prolate asymmetric top, the *a* axis corresponds to the unique axis of the limiting symmetric top. Consequently for a prolate near-symmetric top, the type A band, for which the dipole gradient is parallel to the *a* axis, will show a contour resembling closely the parallel contour of the limiting symmetric top. From what has been said above, it follows that the rotational fine structure of the type A band will show a frequency separation of $2\alpha=B+C$. The B and C type transitions in the prolate asymmetric top both correspond to the perpendicular band of the limiting symmetric top, and the frequency separation of the fine structure, for a near-symmetric top, in these bands will be equal to $2\beta=2A-(B+C)$.

For an oblate near-symmetric top the corresponding expressions are $2\alpha=A+B$ for the type C band, and $2\beta=2C-(A+B)$ for the type A and type B bands.

Using the simulation program it is very easy to verify the above results. As an example we use a harmonic prolate asymmetric top for which the limiting symmetric top rotational constants are 6 and 1 cm^{-1}. The contours calculated by treating the latter as an asymmetric top, *i.e.* using $A=6\ cm^{-1}$, and $B=C=1\ cm^{-1}$, are shown in Figures 2(*a*), 4(*a*) and 7(*a*) for the type A, b and C, respectively. It is clear that the type B and C contours are identical, as they have to be in this limiting case, and that all three show regular fine structure. The fine structure maxima in the type A band are separated by 2 cm^{-1}, while the Q branches in the B and C bands are separated by 10 cm^{-1}, in agreement with the expectations.

Subsequently, the B and C constants were systematically varied, in such a way that (B+C) remains the same.

The values used, together with the values of κ for the resulting rotors, are given in Table 1. The type A, B and C contours that were calculated for these rotors are shown in Figures 2–9. Inspection of these shows that for all three types of bands the fine structure becomes more and more disturbed when the value of $|\kappa|$ decreases, *i.e.* when the asymmetry of the rotor becomes more and more pronounced. For the type A band the fine structure has lost its regular character by the time $|\kappa|$ is lower

Table 1 *Rotational constants* (cm^{-1}) *and* κ *values for the prolate near-symmetric rotors used in the fine structure simulations*

	A	*B*	*C*	κ
1	6.0	1.0	1.0	−1.000
2	6.0	1.05	0.95	−0.960
3	6.0	1.10	0.90	−0.922
4	6.0	1.15	0.85	−0.883
5	6.0	1.20	0.80	−0.846
6	6.0	1.25	0.75	−0.810

than 0.85. For the B and C transitions the regularity persists somewhat longer, but also for these bands it is completely disturbed when $|\kappa|$ is 0.8. It has to be stressed that these simulations have been made under the most favourable conditions. In any real molecule, the harmonic approximation made here is not valid, and the differences between the rotational constants of the upper and lower vibrational states will enhance the disturbances. Moreover, in most real cases an experimental band will also contain contributions from excited state molecules, and this also tends to blur the fine structure. Therefore, for rotors of which $|\kappa|$ is below 0.9, in general no regular rotational fine structure will be experimentally observed.

When a near-symmetric top does show rotational fine structure, the frequency separation, at least when measured on a scale comparable to the relatively low resolution used in this chapter, turns out to be hardly influenced by the degree of asymmetry. This is illustrated in Figures 10–12, in which the fine structure of the type A, B and C bands for different degrees of asymmetry are compared. It is obvious from these figures that the separation of the fine structure is independent of the value of κ. Thus, when regular fine structure is experimentally observed for an asymmetric top, its frequency separation can be used in judging the rotational constants of the top.

This can be illustrated by the conformational analysis of ethyl formate, $HC(O)OCH_2CH_3$.[11] Internal rotation around the (C)O–C(C) bond generates a *trans* and two equivalent *gauche* conformers, in which the C–C bond is *trans* and *gauche* to the (O)C–O(C) bond, respectively. In the vapour phase of this compound, both conformers are present, and in the vapour phase infrared spectrum, several well-separated conformer doublets are observed. Closer inspection of the bands due to one of the conformers reveals that in some cases a weak regular pattern is superposed to their P and R branches, and this pattern is easily recognized as rotational fine structure. As an example, in Figure 13 the R branch of the transition at 781 cm^{-1} is shown. It is evident that from a low resolution spectrum the fine structure maxima cannot be determined with sufficient precision to allow an accurate determination of the rotational constants of the confomer which gives rise to this band. Also, as was discussed above, the fine structure separation is determined by a linear combination of rotational constants.

[11] I. I. Maes, W. A. Herrebout and B. J. van der Veken, to be published.

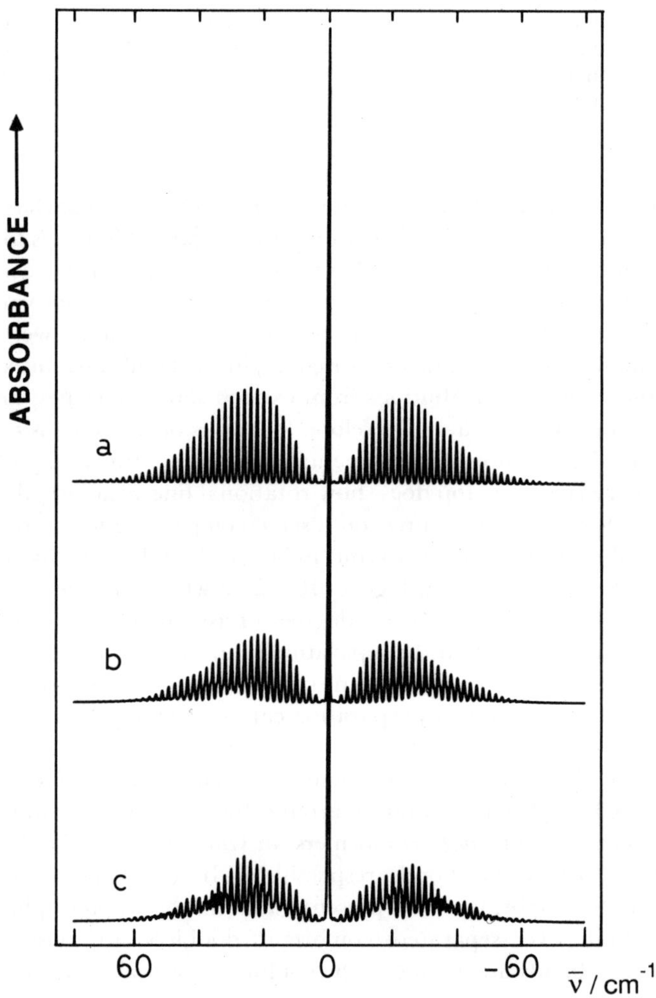

Figure 2 *The rotational fine structure of type A bands of asymmetric top rotors, in the harmonic approximation. The rotational constants, (A, B, C), in* cm^{-1}, *of the rotors are*: (6.0, 1.0, 1.0) (a); (6.0, 1.05, 0.95) (b) *and* (6.0, 1.10, 0.9) (c)

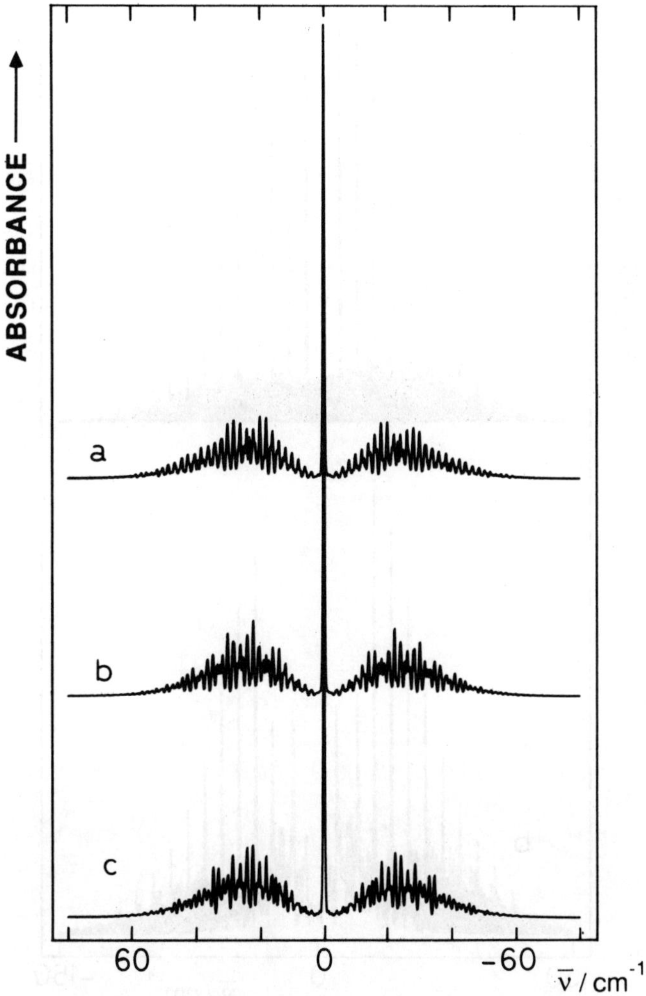

Figure 3 *The rotational fine structure of type A bands of asymmetric top rotors, in the harmonic approximation. The rotational constants, (A, B, C), in* cm^{-1}, *of the rotors are*: (6.0, 1.15, 0.85) (a); (6.0, 1.20, 0.8) (b) *and* (6.0, 1.25, 0.75) (c)

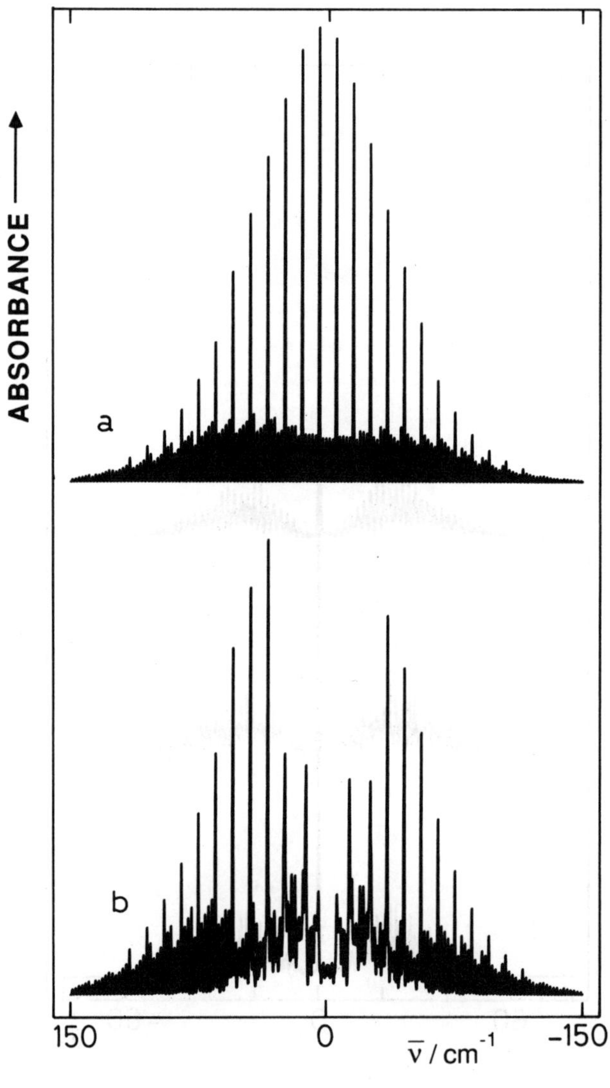

Figure 4 *The rotational fine structure of type B bands of asymmetric top rotors, in the harmonic approximation. The rotational constants, (A, B, C), in* cm^{-1}, *of the rotors are*: (6.0, 1.0, 1.0) (a) *and* (6.0, 1.05, 0.95) (b)

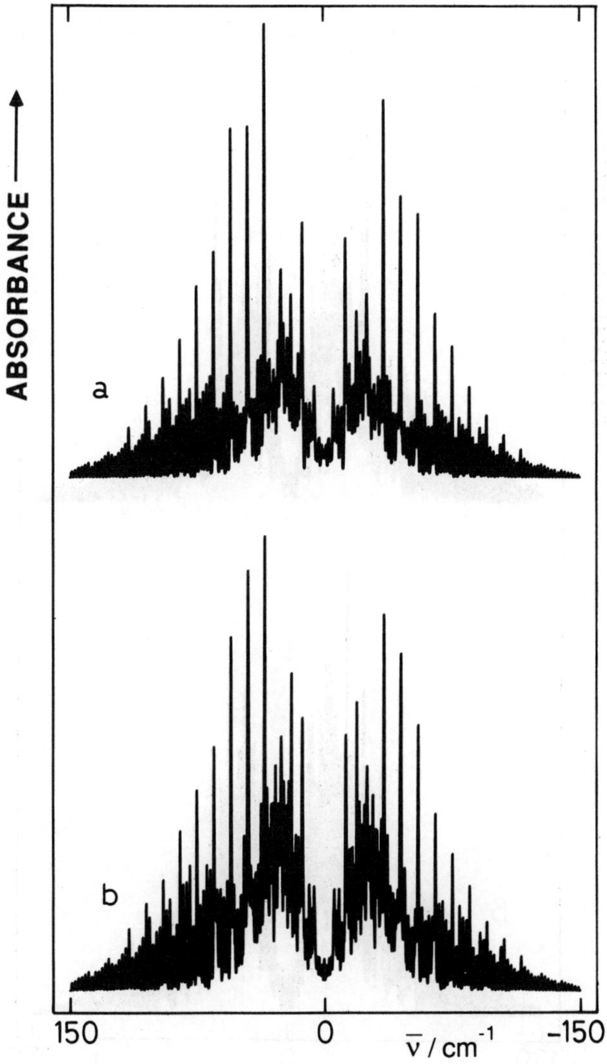

Figure 5 *The rotational fine structure of type B bands of asymmetric top rotors, in the harmonic approximation. The rotational constants (A, B, C), in* cm^{-1}, *of the rotors are*: (6.0, 1.10, 0.9) (a) *and* (6.0, 1.15, 0.85) (b)

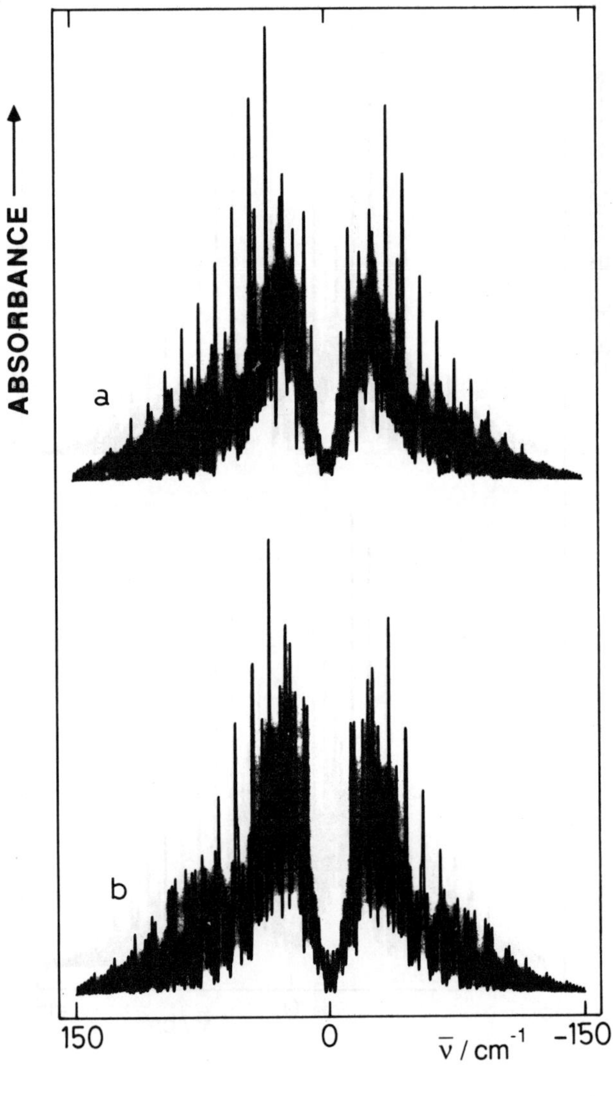

Figure 6 *The rotational fine structure of type B bands of asymmetric top rotors, in the harmonic approximation. The rotational constants (A, B, C), in* cm^{-1}*, of the rotors are*: (6.0, 1.20, 0.80) (a) *and* (6.0, 1.25, 0.75) (b)

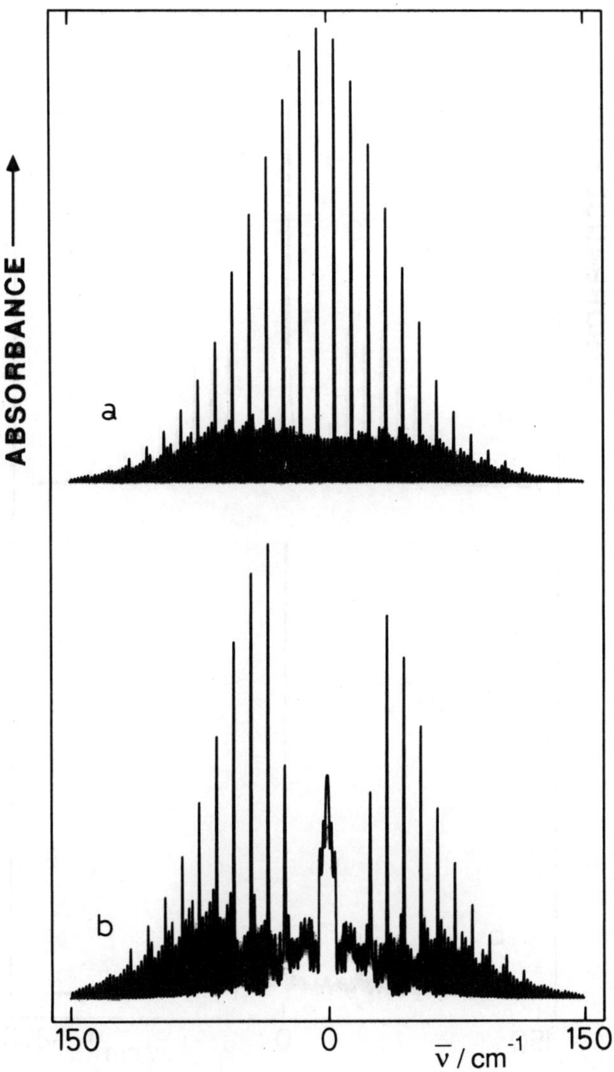

Figure 7 *The rotational fine structure of type C bands of asymmetric top rotors, in the harmonic approximation. The rotational constants (A, B, C), in* cm^{-1}, *of the rotors are*: (6.0, 1.0, 1.0) (a) *and* (6.0, 1.05, 0.95) (b)

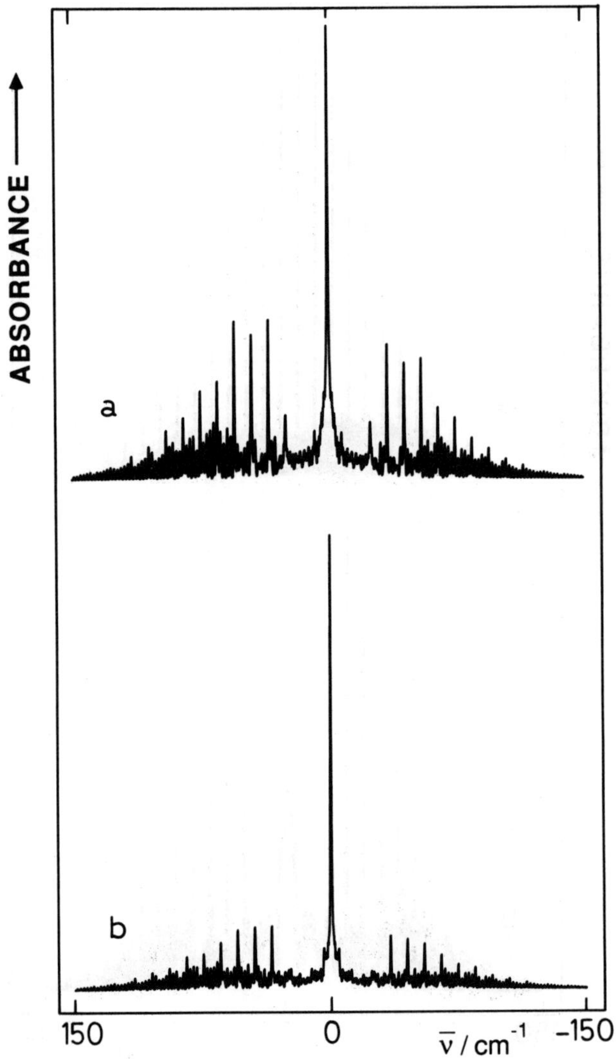

Figure 8 *The rotational fine structure of type C bands of asymmetric top rotors, in the harmonic approximation. The rotational constants (A, B, C), in* cm^{-1}*, of the rotors are*: (6.0, 1.10, 0.9) (a) *and* (6.0, 1.15, 0.85) (b)

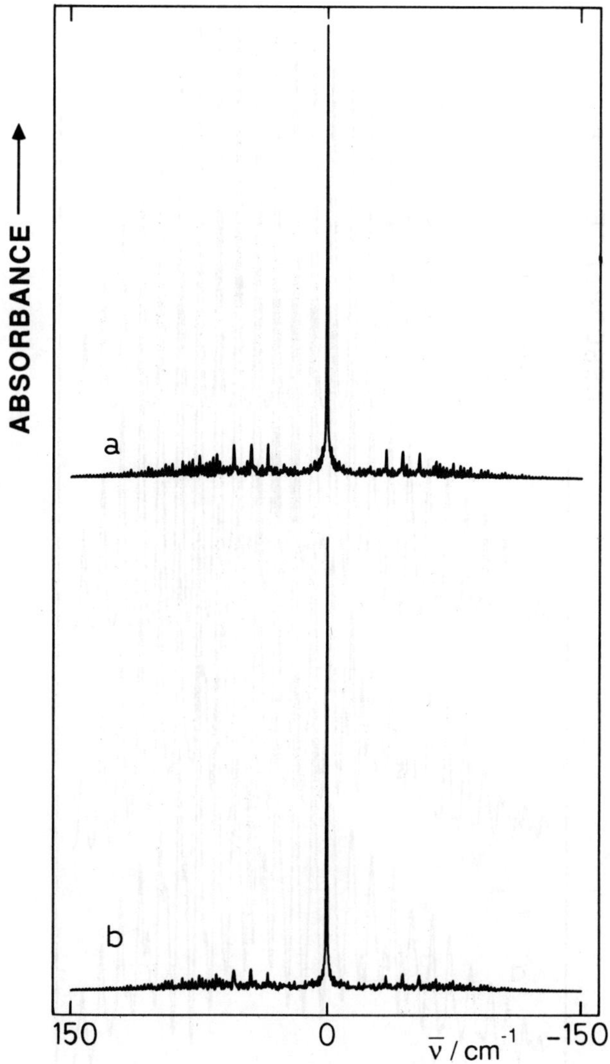

Figure 9 *The rotational fine structure of type C bands of asymmetric top rotors, in the harmonic approximation. The rotational constants (A, B, C), in* cm^{-1}, *of the rotors are*: (6.0, 1.20, 0.8) (a) *and* (6.0, 1.25, 0.75) (b)

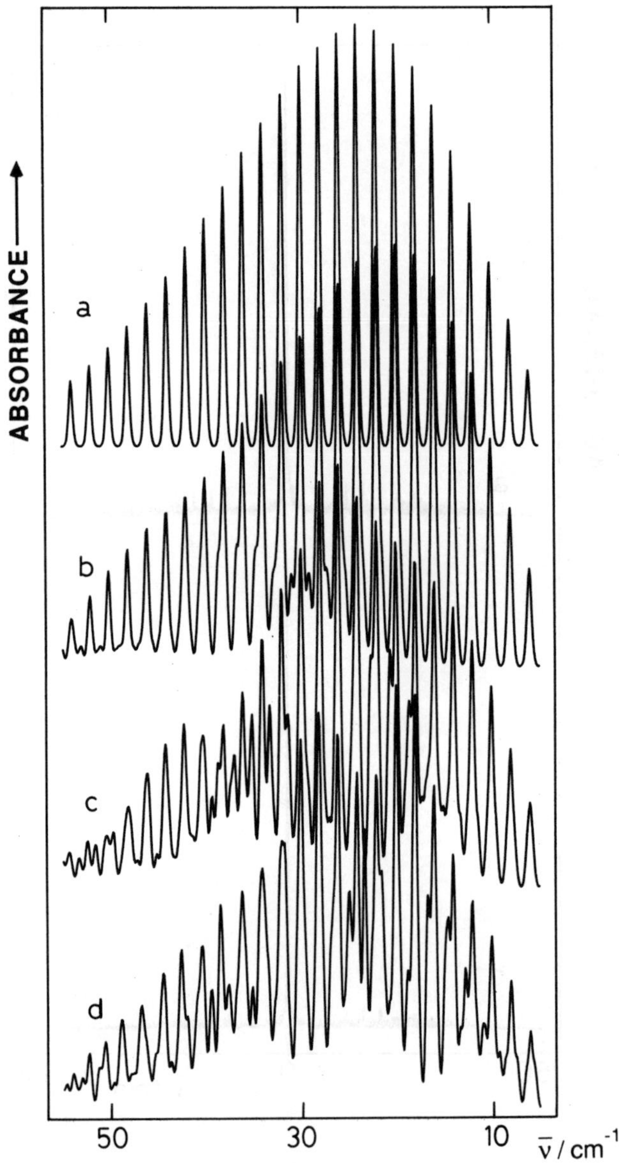

Figure 10 *Comparison of the R-branch type A rotational fine structure. The R-branches shown are those of spectra* a–c *of Fig.* 2 (a–c), *and that of spectrum* (a) *of Fig.* 3 (d)

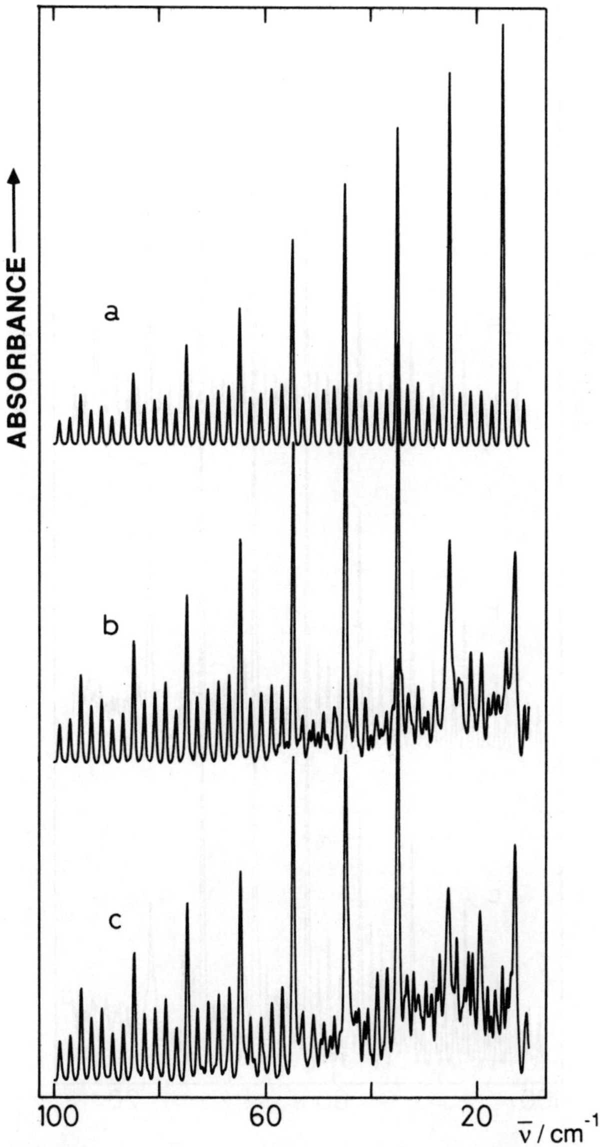

Figure 11 *Comparison of the R-branch type B rotational fine structure. The R-branches shown are those of spectra* (a) *and* (b) *of Fig.* 4 (a,b), *and that of spectrum* (a) *of Fig.* 5 (c)

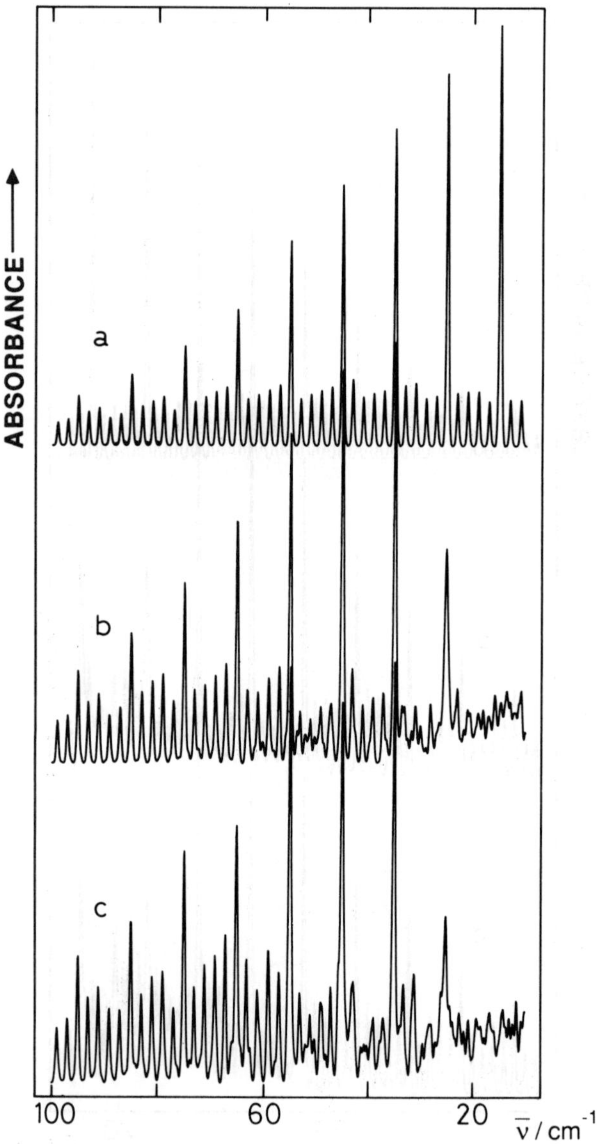

Figure 12 *Comparison of the R-branch type C rotational fine structure. The R-branches shown are those of spectra* (a) *and* (b) *of Fig.* 7, *and that of spectrum* (a) *of Fig.* 8 (c)

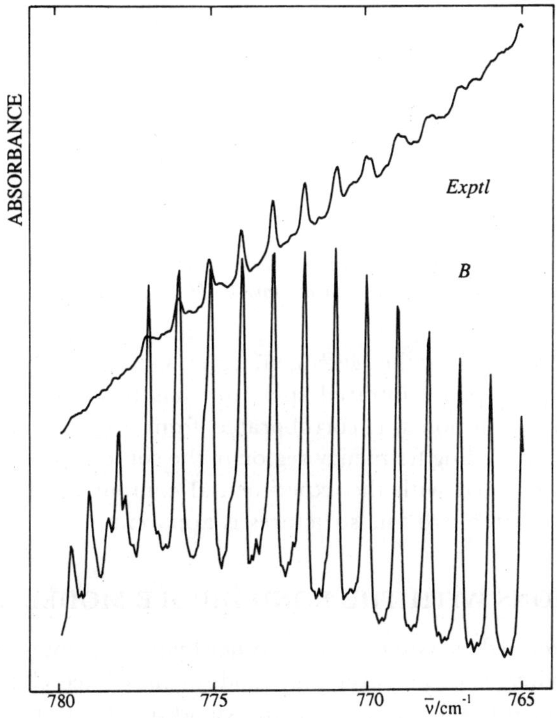

Figure 13 *Rotational fine structure in the P-branch of the band at* 781 cm^{-1} *in the vapour phase infrared spectrum of ethyl formate. Top: experimental spectrum; bottom: simulated type B spectrum*
Published with permission from I. I. Maes, W. A. Herrebout, and B. J. van der Veken, *J. Raman Spectros.*, 1994, **25**, 679.

Thus, even if it is possible to obtain (B + C) from the fine structure of a prolate top, B and C cannot be obtained individually. Therefore, the analysis must follow a different course. For each of the possible conformers a model geometry is set up, and the rotational constants are calculated. Then the required combination of calculated rotational constants are compared with the experimentally observed fine structure separation. For ethyl formate, the rotational constants were calculated from structural data transferred from analogous molecules, and it was found that both the *trans* and the *gauche* conformers are prolate asymmetric tops. The band shown in Figure 13 is pronouncedly of type B, and therefore it was assumed that the observed fine structure is due to the type B contribution, so that the separation must be equal to 2A − (B + C). For the *trans* conformer, κ was calculated to be −0.958, and 2A − (B + C) was calculated to be equal to 1.01 cm^{-1}. For the *gauche* conformer the values are −0.561 and 0.32 cm^{-1}, respectively. The calculated values of κ suggest strongly that for the *gauche* conformer no regular fine structure can be observed, while for the *trans* conformer it in principle must be observable. Also, the experimental separation neatly agrees with the value of B + C predicted for the *trans* conformer, while it differs strongly

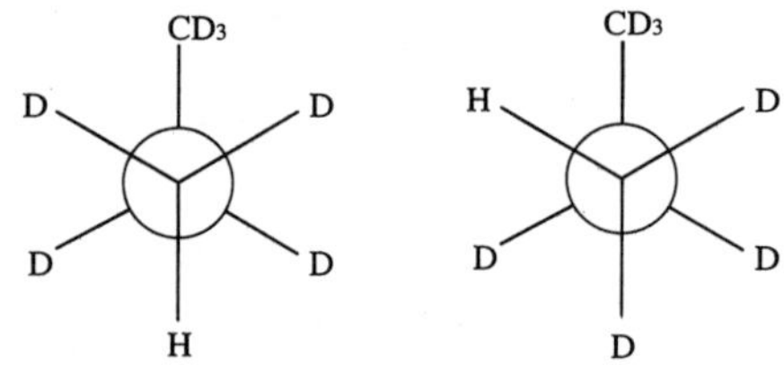

Figure 14 *Newman projections of the possible orientations of the isolated C–H bond in* [$^2H_7$] *propane,* $CHD_2CD_2CD_3$. (a) *in-plane C–H* bond; (b) *out-of-plane C–H bond*

from the value predicted for the *gauche* conformer. Consequently, the band at 781 cm^{-1} in the vapour phase infrared spectrum has to be assigned to the *trans* conformer. This conclusion was corroborated from a simulation of the type B contour: the corresponding frequency region of the contour predicted for the *trans* conformer is superposed with the experimental contour in Figure 13, and the agreement between the two fine structures is apparent.

9 SIMULATIONS WITH THE BOND DIPOLE MODEL (BDM)

A C–H bond is termed 'isolated' when no other hydrogen atoms are connected to the carbon atom involved. The stretching mode of an isolated C–H bond is known to be extremely characteristic,[12] to such an extend that its stretching frequency is an accurate measure of the bond length.[13] The high degree of localization of the stretching mode in this type of bond suggests that the dipole gradient $(\partial\mu/\partial Q)_0$ for this vibration must be accurately predicted by the bond dipole model (BDM). This is confirmed for a number of cases, some of which will be discussed below. It must be noted, however, that in a later paragraph isolated C–H stretches will be discussed for which the dipole gradient diverges strongly from the BDM prediction: such deviations are induced by electron-rich atoms in the immediate vicinity of the isolated C–H bond. The present examples, of course, do not belong to the class of molecules were such effects are likely to occur.

The first example in which BDM dipole gradients are used is the simulation of the isolated C–H stretching region of [$^2H_7$] propane, $CD_3CD_2CHD_2$, in which a single hydrogen atom is attached to a methyl carbon. In Figure 14 the Newman projections are given for the possible orientations of the isolated C–H bond. The latter can be oriented *trans* to the CD_3 grouping, which we will call the in-plane orientation. The C–H bond can also occur in two equivalent *gauche* positions, and these will be called the out-of-plane orientation. The two orientations are connected *via* the internal rotation around the $(D_2H)C–C(D_2)$ bond, which is hindered by a finite barrier. Therefore, in a room temperature sample, both orientations will be present. The difference in free enthalpy between the molecules having the different orientations will be very small, and the relative amounts in a vapour phase sample will be dictated by the statistical weights of the two forms:

[12] D. C. McKean, *Chem. Soc. Rev.*, 1978, **7**, 399.
[13] D. C. McKean, *J. Mol. Struct.*, 1984, **113**, 251.

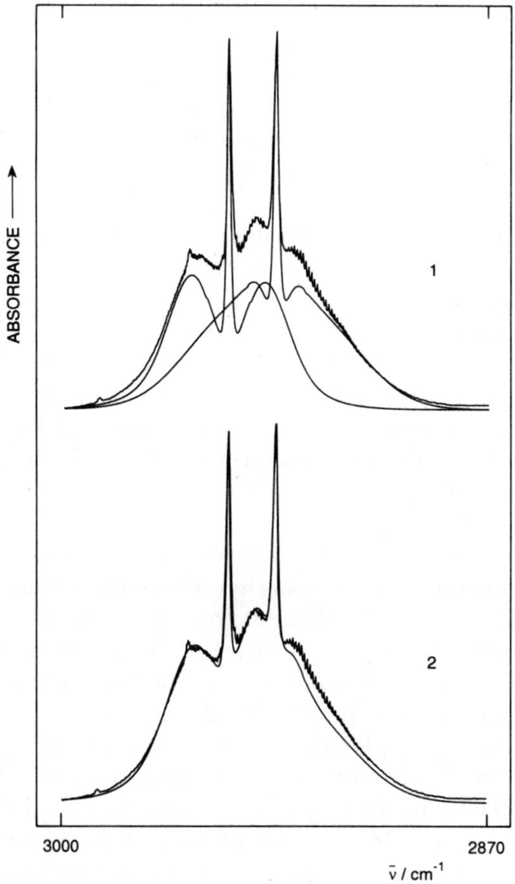

Figure 15 *The isolated C–H stretches in* [$^2H_7$] *propane.* 1, *Comparison of the experimental contour with the individual BDM hybrids* (*not drawn to scale*); 2, *comparison of the experimental contour with the calculated sum contour*

because two equivalent out-of-plane forms exist, the statistical weight of this form is two, compared to one for the in-plane form. The two forms have their C–H bond surrounded by different environments, and will show slightly different C–H stretching frequencies. In the vapour phase infrared spectrum two C–H stretches are observed, at 2950 and 2936 cm^{-1}, respectively. The experimental contour is shown in Figure 15. The simulation of this contour serves two purposes. In the first place the simulation will confirm the assignments of the two transitions to the different forms, which had previously been made based on other grounds,[12] and it also allows the relative intensities to be obtained: because of the overlap between the two transitions these relative intensities are difficult to obtain directly from the vapour phase spectrum. The C–H stretching region was simulated[14] in the

[14] B. J. van der Veken, A. Grupping and D. C. McKean, to be published.

harmonic approximation. For both isolated C–H stretches the dipole gradient was taken parallel to its C–H bond, in agreement with the BDM model. It was found that the simulation in which the in-plane stretch is assigned to the high frequency component, and the out-of-plane stretch to the low frequency component was far superior to the simulation in which the assignment was reversed. The results for the former are shown in Figure 15. In the upper half (not shown with the correct relative intensities) the experimental contour is superposed with the optimized individual hybrids, while in the lower half the experimental contour is superposed with the resulting sum contour. The nature of the weak Q branch, present near the R branch maximum of the 2950 cm^{-1} transition, is not known, and no attempts were made to simulate it.

It can be seen in Figure 15 that the agreement of the calculated with the experimental contour is quite good; only in the region of the P branch of the low frequency transition is the calculated contour somewhat less intense than the experimental one. This is thought to be primarily owing to the presence, inferred from the hump on the experimental P branch, of one of the other isomers of $[^2H_7]$propane, CD_3CHDCD_3.

These simulations, as expected, confirmed the previous assignments. The intensity ratio of in-plane to out-of-plane hybrids for the simulation in Figure 15 is 0.76 ± 0.07. This is significantly different from the ratio 0.5 expected from the statistical weights of the two forms, and suggests that the extinction coefficients for the two C–H stretches are different. This is in accord with the studies of Kindness *et al.*,[15] who found that the intensity ratio for CH_{eq} to CH_{ax} in *c*-$C_6D_{11}H$ was 1.375. Thus a C–H *trans* to a carbon is enhanced relative to a C–H *trans* to a hydrogen.

A similar problem is encountered in $[^2H_9]$butane. For this compound the isotopic derivatives with the hydrogen atom in the methyl and the one with the hydrogen atom in the methylene grouping have both been synthesized, and their spectra are available. Compared to propane, things in butane are complicated by the presence of two conformers, the *trans* and *gauche* conformer, in which the C(1)–C(2) and C(3)–C(4) carbon–carbon bonds are *trans* and *gauche* to each other, respectively. The possible positions of the hydrogen atom are given in Figure 16. For the derivative with the methyl hydrogen it can be seen that two possible orientations, T1 and T2, arise in the *trans* conformer, and three different orientations, G1, G2 and G3, can occur in the *gauche* conformer. Thus, for this derivative a total of five different C–H stretches must be expected. The experimental spectrum is shown in Figure 17. In this spectrum, Q branches are present at 2950, 2937 and 2933.8 cm^{-1}. This spectrum was simulated in the harmonic approximation, using BDM dipole gradients for each isolated C–H stretch.[16] The systematic analysis led to an unambiguous assignment at least for the dominant transitions. The two best theoretical sum contours are compared with the experimental one in Figure 17. In both these simulations, the intense high frequency component, at 2950 cm^{-1}, was simulated using a 0.75 : 0.25 mixture of the T1 and G3 hybrids, while the band at 2937 cm^{-1} was simulated using the T2

[15] A. Kindness, D. C. McKean and D. Stewart, *J. Mol. Struct.*, 1990, **224**, 363.

[16] B. J. van der Veken, R. O. Odeurs, A. Brown, D. C. McKean and A. R. Morrison, *J. Mol. Struct.*, 1986, **147**, 57.

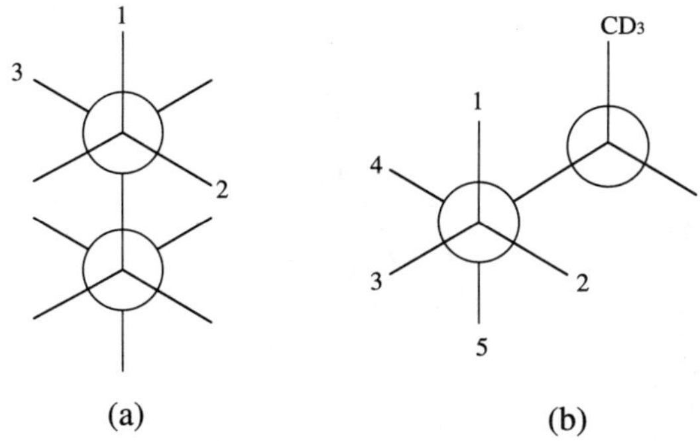

Figure 16 *Newman projections of the possible orientations of the isolated C–H bond in* [$^2H_9$] butane. (a) trans *conformer*; (b) gauche *conformer*

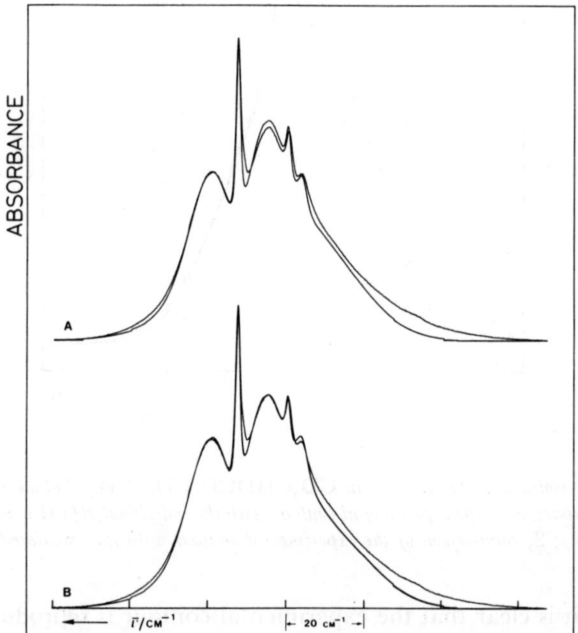

Figure 17 *The isolated C–H stretches in* $CHD_2CD_2CD_2CD_3$, [$^2H_9$] *butane* (*methyl–H*). *In both cases the experimental contour is compared with a calculated sum contour. The composition of the calculated sum contours is described in the text*
Published with permission from B. J. van der Veken, R. O. Odeurs, A. Brown, D. C. McKean, and A. R. Morrison, *J. Mol. Struct.*, 1986, **147**, 57.

conformer. For simulation A, the low frequency component was taken to be composed of equal parts of the G1 and G2 hybrids, and for simulation B this band was simulated using the G2 hybrid only. The agreement with experiment of simulation A is somewhat better than that of B, which led the authors to conclude that the G1 contour is presumably accidentally degenerate, and unobserved because of its relative weakness, with either the 2950 or the 2937 cm^{-1} bands.

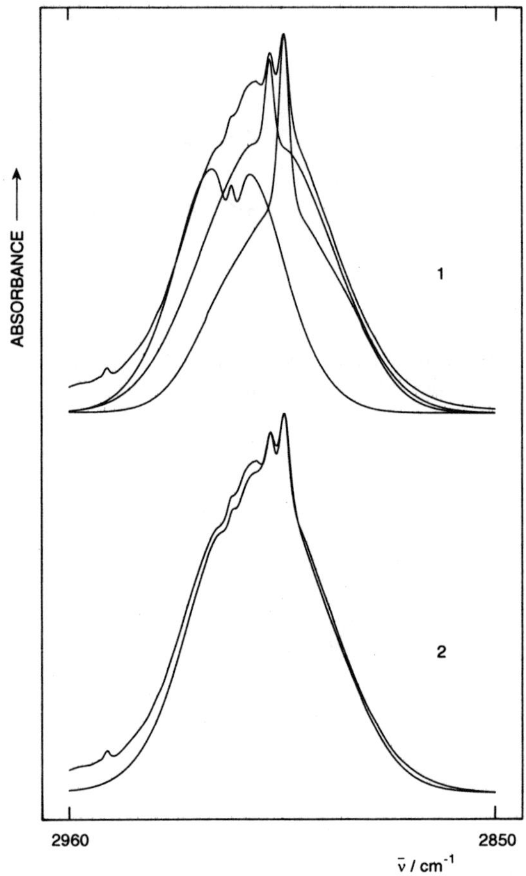

Figure 18 *The isolated C–H stretches in* $CD_3CHDCD_2CD_3$, [$^2H_9$] *butane* (*methylene–H*). 1, *Comparison of the experimental contour with the individual BDM hybrids (not drawn to scale)*; 2, *comparison of the experimental contour with the calculated sum contour*

Altogether, it is clear that the experimental contour is reproduced very well, except in the low frequency region. This deviation is caused mainly by the weak band near 2910 cm^{-1} in the experimental spectrum, which, as will be seen in the next paragraph, is due to the isotopic derivative with the methylene hydrogen.

For the isotopic derivative with the methylene hydrogen, from Figure 16 a simpler structure of the C–H stretching region can be anticipated, because only one *trans*, T3, and two *gauche* hybrids, G4 and G5, are expected to contribute. The C–H stretching region of this derivative is shown in Figure 18. Although a pattern with strongly overlapping transitions are present, band centres are identified easily at 2918, 2907 and 2904 cm^{-1}. The weak band at 2950 cm^{-1} is due to the other isotopic derivative, present as an impurity. Also this contour was analysed in the harmonic approximation, using BDM dipole gradients.[14] The results are given in Figure 18: the high frequency component was simulated using the T3 hybrid, the

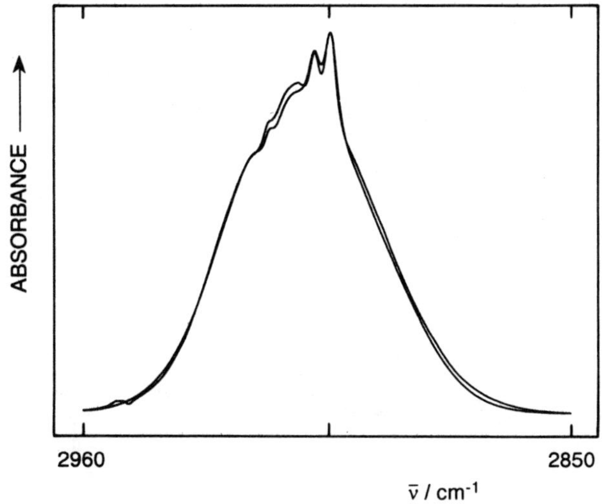

Figure 19 *Comparison of the calculated sum contour for the C–H stretches in* [$^2H_9$] *butane (methylene–H) with the experimental contour corrected for* [$^2H_9$] *butane* (methyl–H)

bands at 2907 and 2904 cm^{-1} were simulated using the G3 and G4 hybrids, respectively. From the comparison of the sum contour with the experimental one in the lower half of Figure 18, it can be seen that the agreement in the low frequency region is excellent, but that in the high frequency region the theoretical contour is consistently less intense than the experimental contour. Because, as can be seen in Figure 17, the isotopic impurity at 2950 cm^{-1} has considerable intensity in virtually the whole high frequency region of Figure 18, the experimental contour was corrected by subtracting the scaled spectrum of the methyl–hydrogen derivative. The comparison of the corrected experimental contour with the theoretical sum contour is shown in Figure 19. It is now clear that an excellent agreement over the whole contour is achieved.

The simulations for both isotopes of [$^2H_9$] butane again support the assignments of the isolated C–H stretches that had been made on different grounds. Especially for the methylene–hydrogen derivative it will be appreciated that from the simulations the relative intensities of the different components can be obtained, which were totally inaccessible from the experimental contour.

Because of their dependence on rotational constants, the vapour phase infrared contours of asymmetric tops must contain structural information on the molecules in which they arise. In view of the complexity of the experimental spectra of mid-size asymmetric top molecules, and as a consequence of the low resolution with which they are investigated here, it is evident that it is impossible to derive from the experimental contours the rotational constants with a precision that allows a structure to be determined with the accuracy we have used from other structural techniques. There is, however, a branch of structural chemistry that does benefit from the type of analysis applied here. This branch is conformational analysis. The reason for the success of contour simulation in conformational

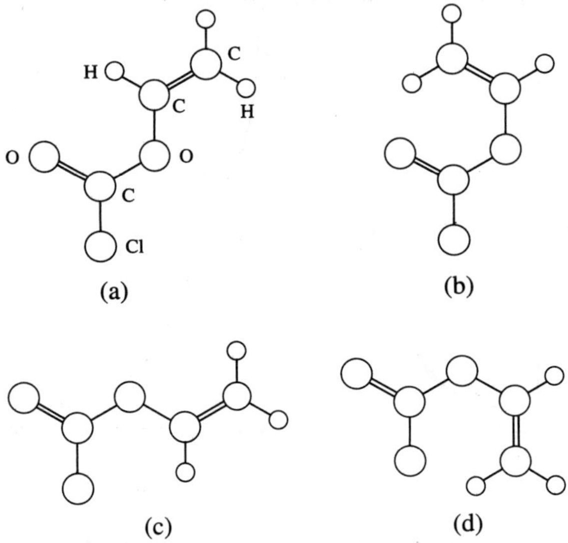

Figure 20 *The possible conformers of vinyl chloroformate.* (a) *s*-cis,trans *conformer*; (b) *s*-cis,cis *conformer*; (c) *s*-trans, trans *conformer*; (d) *s*-trans,cis *conformer*

analysis lies in the fact that different conformers very often have widely different rotational constants. In such cases, the different conformers have different low resolution contours, and these differences can be exploited either to identify the conformer present in the vapour phase, or to assign components in conformer multiplets to specific conformers.

As a first example of a conformational assignment, we will discuss vinyl chloroformate,[17] ClC(O)OCH=CH_2. In this compound, conformers are generated by rotation around the (O=)C–O(C) and (C)O–C(=C) bonds. The conjugation in this molecule will cause all conformers to be planar. Projections of the possible conformers are shown in Figure 20. In agreement with accepted practice, we will indicate the confomers due to rotation around the (O=)C–O(C) bond as s-*cis* and s-*trans* when the (C)O–C(=C) is *cis* or *trans* to the carbonyl bond, respectively. The orientations of the C=C bond will be described as *cis* or *trans*, depending on its orientation relative to the (O=)C–O(C) bond. From *ab initio* calculations it was shown that only the s-*cis,trans* and s-*cis,cis* conformers have reasonable energies, and, therefore, only these two conformers were considered in the analysis.[17]

The vapour phase infrared spectra of vinyl chloroformate suggest that a single conformer is present. In the spectra, the C=O and C=C stretches appear well-separated from other bands, and show well-defined P, Q, R structures. Also because of their characteristics, these bands were selected for simulation. All calculations were made in the harmonic approximation, and at the outset BDM dipole gradients were used for both vibrations. In Figures 21 and 22, the experimental ν(C=O) contour is compared to the predicted ones. Figure 21 shows

[17] B.J. van der Veken, Jie Lin and J. R. Durig, *Struct. Chem.*, 1993, **4**, 167.

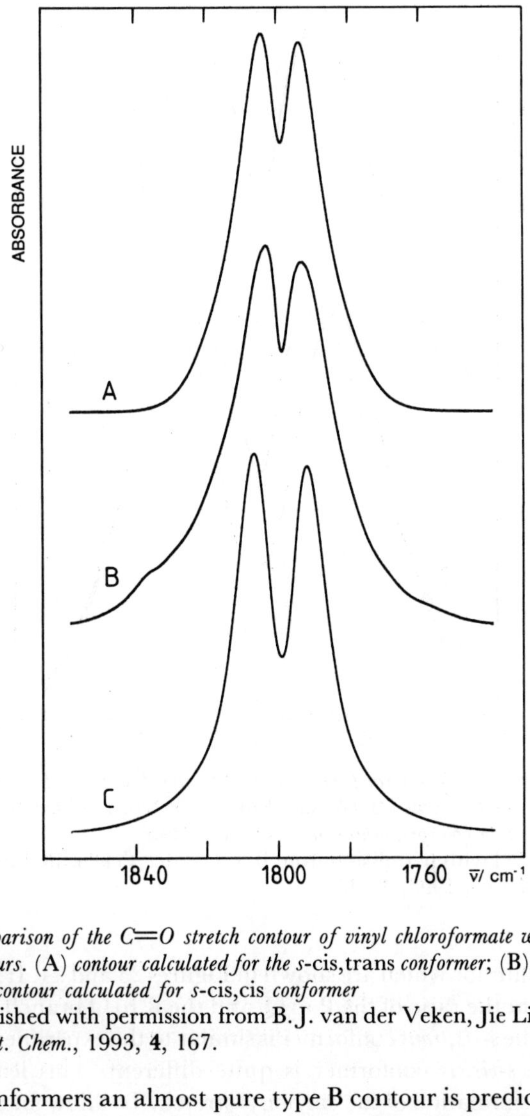

Figure 21 *Comparison of the C=O stretch contour of vinyl chloroformate with calculated BDM contours.* (A) *contour calculated for the s-*cis,trans *conformer*; (B) *experimental contour*; (C) *contour calculated for s-*cis,cis *conformer*
Published with permission from B. J. van der Veken, Jie Lin, and J. R. Durig, *Struct. Chem.*, 1993, **4**, 167.

that for both conformers an almost pure type B contour is predicted, and that the experimental contour also is a virtually pure type B band. In Figure 22 the region near the band centre is compared in more detail. It can be seen that the frequency separation between the two maxima, called the PQ,QR separation for this type of band, for the s-*cis,trans* hybrid closely matches the experimental value, while the separation found in the s-*cis,cis* hybrid is significantly larger. These simulations are an example where it is not the overall character of the contour, but another characteristic, in this case the PQ, QR separation, that allows the acceptance of one model, the s-*cis,trans* conformer, and the rejection of another, the s-*cis,cis* conformer. Therefore, from the simulation of the carbonyl stretching it was concluded that in the vapour phase vinyl chloroformate has the s-*cis,trans* conformation. This conclusion was corroborated from the simulations of the C=C

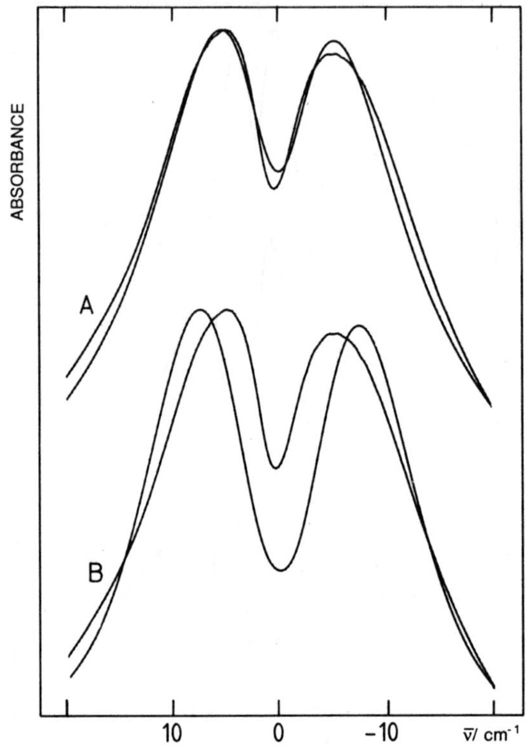

Figure 22 *Comparison of the central part of the C=O stretch contour of vinyl chloroformate with calculated BDM contours.* (A) *superposition of experimental and s-*cis,trans *contour*; (B) *superposition of experimental and s-*cis,cis *contour*
Published with permission from B. J. van der Veken, Jie Lin, and J. R. Durig, *Struct. Chem.*, 1993, **4**, 167.

stretching, the results for which are shown in Figures 23 and 24. It is easy to deduce from Figure 23 that the ratio of the P to Q and the R to Q branch intensities of the BDM hybrid for the s-*cis,trans* conformer is similar to the experimental value, while this ratio for the s-*cis,cis* conformer is quite different. This leads to the same conclusion as the simulation of the carbonyl stretch. To allow for a possible failure of the BDM, the C=C stretch contour, which is predominantly type A, was also compared with the pure type A transitions of the two conformers. The results of this are given in Figure 24, and it is clear that the PR separation for the s-*cis,trans* type A band agrees with the experimental value, while that for the s-*cis,cis* type A band does not. So, also from this comparison, it had to be concluded that vinyl chloroformate has the s-*cis,trans* conformation.

An analogous conformational problem is found in chloromethyl formate, $HC(O)OCH_2Cl$. The internal rotation around the (O=)C–O(C) bond again generates s-*cis* and s-*trans* conformers, but now the internal rotation around the (C)O–C(Cl) bond gives rise to *trans* and *gauche* conformers, in which the C–Cl bond is *trans* or *gauche* to the (O)=C–O(C) bond, respectively. Also for this

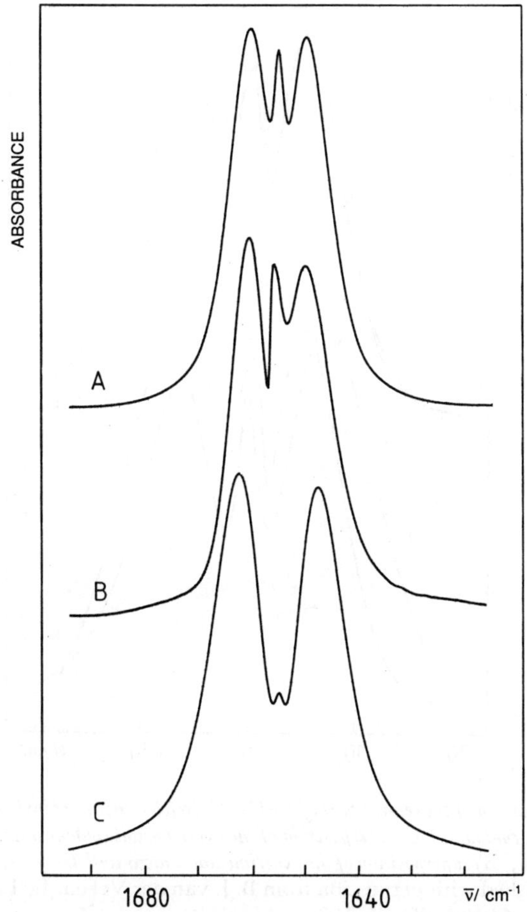

Figure 23 *Comparison of the C═C stretch contour of vinyl chloroformate with calculated BDM contours.* (A) *contour calculated for the s-*cis,trans *conformer;* (B) *experimental contour;* (C) *contour calculated for the s-*cis,cis *conformer*
Published with permission from B. J. van der Veken, Jie Lin, and J. R. Durig, *Struct. Chem.*, 1993, **4**, 167.

compound the vapour phase spectra indicate that only one conformer is present.[18] The formate C–H stretch of this compound is an isolated C–H stretch, and in the isotopic derivative $HC(O)OCD_2Cl$ it appears well separated from other bands. The results of the simulation of this band,[18] in the harmonic approximation, and using BDM dipole gradients, are shown in Figure 25. It is immediately obvious that the hybrids for the s-*cis,trans* and s-*trans,trans* conformers are totally incompatible with the experimental contour. The s-*cis,gauche* and s-*trans,gauche* hybrids show P and R branches of the required relative intensity. Closer inspection shows that the PR separation of the s-*trans,gauche* hybrid is significantly smaller

[18] F. Daeyaert and B. J. van der Veken, *J. Mol. Struct.*, 1989, **213**, 97.

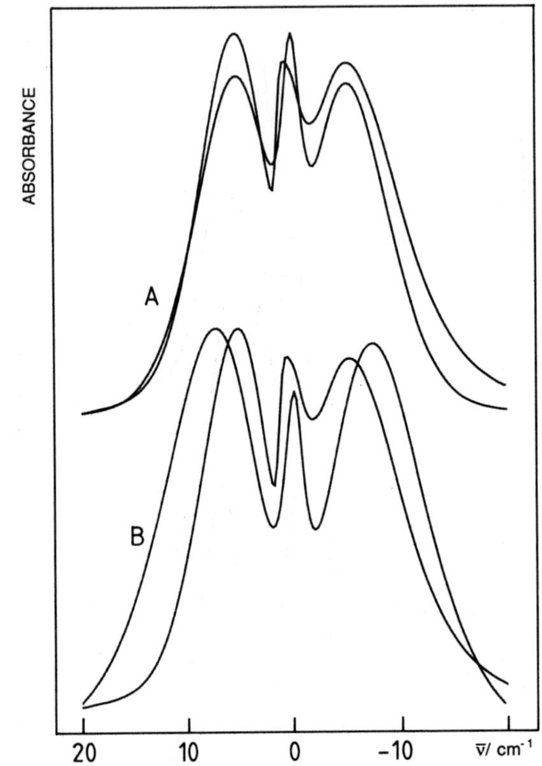

Figure 24 *Comparison of the central part of the C═C stretch contour of vinyl chloroformate with pure type A contours.* (A) *superposition of the experimental contour with the s-*cis,trans *type A contour;* (B) *superposition of the experimental contour with the s-*cis,cis *pure type A contour* Published with permission from B. J. van der Veken, Jie Lin, and J. R. Durig, *Struct. Chem.*, 1993, **4**, 167.

than the experimental value, and that this separation for the s-*cis,gauche* hybrid agrees much better. Therefore, from the simulations of this C–H stretching, it follows that chloromethyl formate in the vapour phase occurs as the s-*cis,gauche* conformer. This assignment was checked using the contour of the symmetric CD_2 stretching in the same isotopic derivative. The resulting separations are compared with the experimental contour in Figure 26. For this vibration, not only the s-*cis,trans* and s-*trans,trans* but also the s-*trans,gauche* hybrid can be rejected at once. On the other hand, the s-*cis,gauche* hybrid reasonably reproduces the experimental contour. The agreement is less than perfect, however; the reason for this must be the less accurate prediction of the dipole gradient by the BDM due to the presence of the electron-rich chlorine and oxygen atoms. This effect should also affect the dipole gradient for the formate C–H stretch discussed above, but in that case both substituents on the carbon atom are oxygen atoms, and their reorienting influences presumably all but cancel.

A completely analogous conformational problem has to be solved in chloromethyl

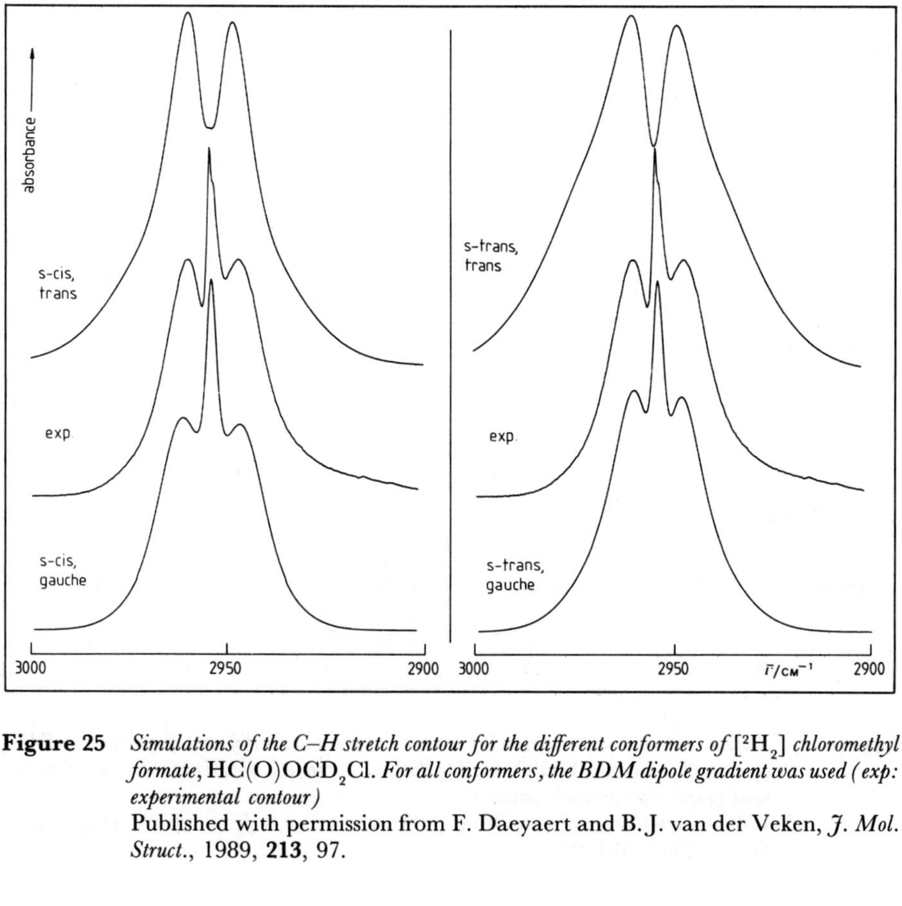

Figure 25 *Simulations of the C–H stretch contour for the different conformers of* [$^2H_2$] *chloromethyl formate*, HC(O)OCD$_2$Cl. *For all conformers, the BDM dipole gradient was used (exp: experimental contour)*
Published with permission from F. Daeyaert and B.J. van der Veken, *J. Mol. Struct.*, 1989, **213**, 97.

fluoroformate, FC(O)OCH$_2$Cl. For this compound, the carbonyl stretch and the symmetric CH$_2$ stretch were used for simulation.[19] Again all four possible conformers were tested, in the harmonic approximation and using BDM dipole gradients. The experimental carbonyl stretching contour, as can be seen in Figure 27, is rather disturbed but still exhibits a pronounced type B contour. Only the s-*cis,gauche* and s-*cis,trans* conformers predict a sufficiently type B contour. The central dip in the s-*cis,trans* hybrid and its relatively high intensity in the wings make the s-*cis,trans* confomer a less likely candidate, so that for this compound it was concluded that in the vapour phase it is composed of the s-*cis,gauche* conformer. It is clear, however, that the above evidence for the identity of the conformer is less convincing than for the previous examples. Fortunately, from the simulations of the symmetric CH$_2$ stretching, shown in Figure 28, the s-*cis,trans* can be definitely ruled out. Thus, by combining the results for two different vibrations, an unambiguous identification of the vapour phase conformer is possible.

[19] F. Daeyaert and B.J. van der Veken, *Spectrochim. Acta*, 1989, **45A**, 993.

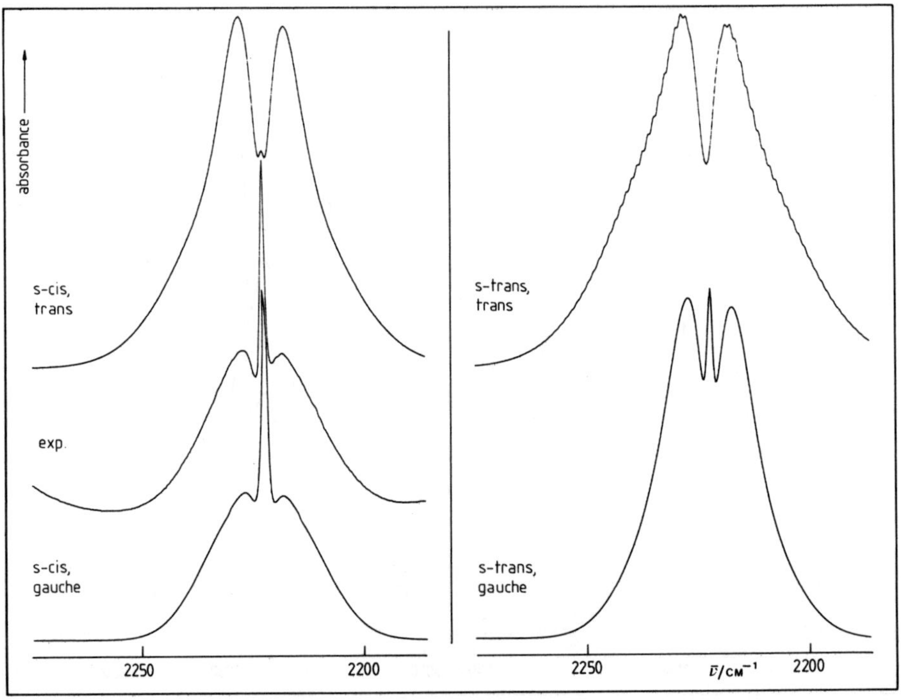

Figure 26 *Simulations of the symmetric* CD_2 *stretch contour for the different conformers of* $[^2H_2]$ *chloromethyl formate,* $HC(O)OCD_2$*. For all conformers, the BDM dipole gradient was used (exp: experimental contour)*
Published with permission from F. Daeyaert and B. J. van der Veken, *J. Mol. Struct.*, 1989, **213**, 97.

10 CONFORMATIONAL APPLICATIONS USING NON-BDM GRADIENTS

For many vibrations in asymmetric top molecules it will be found that the dipole gradient cannot be predicted from the bond dipole model (BDM). When the direction of the gradient cannot be reliably predicted, it may be attempted to deduce the composition of an experimental band by systematically comparing with all possible hybrids. As the sum of the squares of the direction cosines of the dipole gradient must be equal to one, testing all possible hybrids amounts to a two-dimensional search. It will be shown in a later example that in spite of the complexity, such two dimensional searches can be used in conformational analysis. We will begin, however, with a number of examples in which, either by symmetry or by chemical intuition, the search is restricted to one dimension, *i.e.* only one parameter needs to be varied to scan all possible orientations of the dipole gradient.

A first example in which one-dimensional searches are used is the conformational

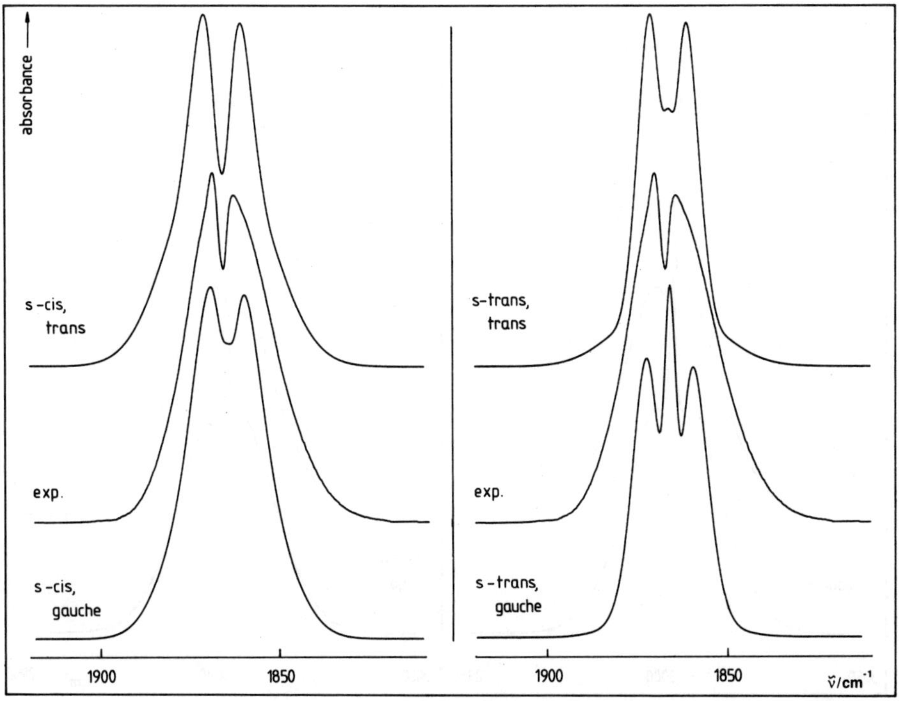

Figure 27 *Simulations of the C═O* stretch contour for the different conformers of chloromethyl fluoroformate, $FC(O)CH_2Cl$, *using BDM dipole gradients (exp: experimental contour)*
Published with permission from F. Daeyaert and B.J. van der Veken, *Spectrochim. Acta, Part A*, 1989, **45**, 993.

analysis of methyl nitrite, CH_3ONO.[20] It has long been established, for instance from the occurrence of a number of conformer doublets in the infrared spectrum, that in the vapour phase this compound is composed of two conformers. The structures of these have been established by microwave spectroscopy[21] as a *cis* and a *trans* conformer, projections of which are shown in Figure 29. The assignments of the components of conformer doublets in the vibrational spectra by different authors have not always been consistent, and the best evidence available at the time the simulations were made was, in spite of its ingenuity, only indirect, as it was based on photolysis experiments in an argon matrix.[22]

The N═O stretch in this compound shows a well-separated conformational doublet, and at the outset it was attempted to simulate the contours using BDM hybrids. These attempts failed, showing that the dipole gradient for this stretching mode is not parallel to the N═O bond. Both conformers have C_s symmetry, with the N═O bond lying in the plane of symmetry. Therefore, by symmetry, the

[20] B.J. van der Veken, R. Maas, G.A. Guirgis, H.D. Stidham, T.G. Sheehan and J.R. Durig, *J. Phys. Chem.*, 1990, **94**, 4029.
[21] P.H. Turner, M.J. Corkill and A.P. Cox, *J. Phys. Chem.*, 1979, **83**, 1473.
[22] M.E. Jacox and F.L. Rook, *J. Phys. Chem.*, 1982, **86**, 2899.

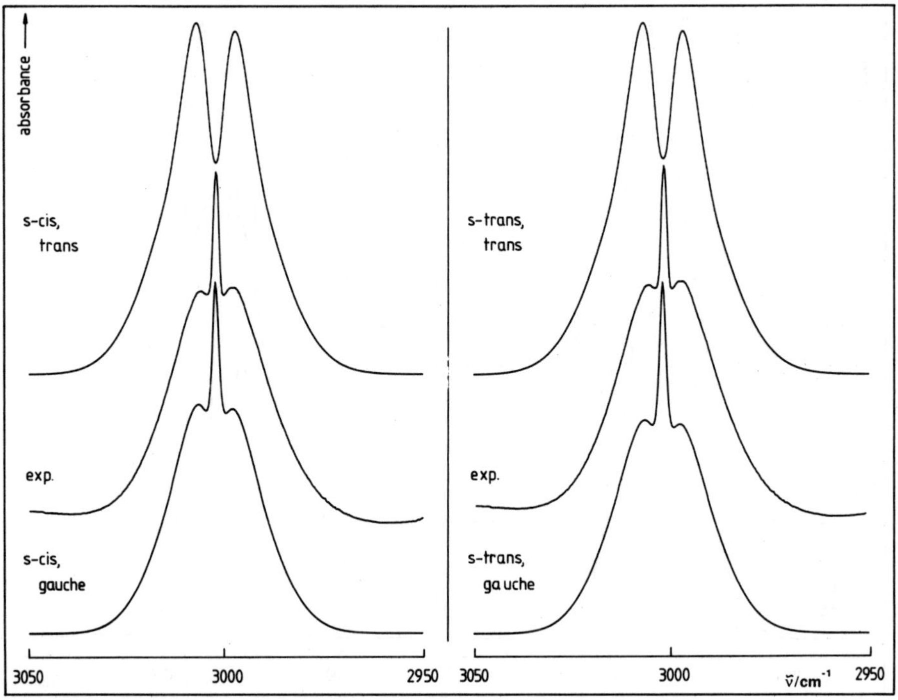

Figure 28 *Simulations of the CH_2 stretch contour for the different conformers of chloromethyl fluoroformate,* FC(O)OCH$_2$Cl, *using BDM dipole gradients (exp: experimental contour)* Published with permission from F. Daeyaert and B.J. van der Veken, *Spectrochim. Acta, Part A*, 1989, **45**, 993.

dipole gradient must also lie in this plane. As the latter for both conformers contains the *a* and *b* principal axes, the N=O stretch must be an A/B hybrid. All possible hybrids are therefore generated by sweeping the gradient across the *a,b*-plane: this can be described by a single variable, for instance the angle between the dipole gradient and the N=O bond.

In cases where the different conformers of a molecule have similar rotational constants, the pure type transitions of different conformers will be very similar, and by relaxing the orientation of the dipole gradient, each of the conformers will allow, with similar success, the simulation of each component of a conformational multiplet. For such cases, it is very difficult to draw conformational conclusions. For methyl nitrite, however, the *cis* and *trans* conformer have rather different rotational constants, and it can be seen in Figures 30 and 31 that the one-dimensional searches allow an unambiguous assignment. The simulations of the high frequency N=O stretch, at 1677 cm^{-1}, are shown in Figure 30. In the left hand side of the figure, *trans* hybrids for different orientations of the dipole gradient are compared with the experimental contour; in the right hand side *cis* hybrids are shown. Because of the presence of an excited state, the Q branch is split into a doublet. To account for this, the calculated contours were constructed by

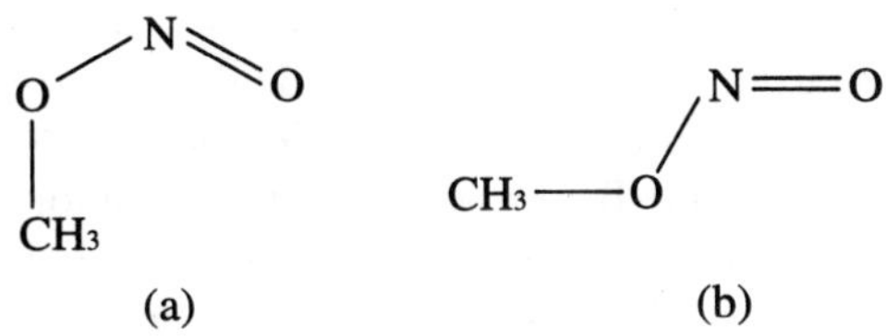

Figure 29 *The* cis (a) *and* trans (b) *conformers of methyl nitrite*

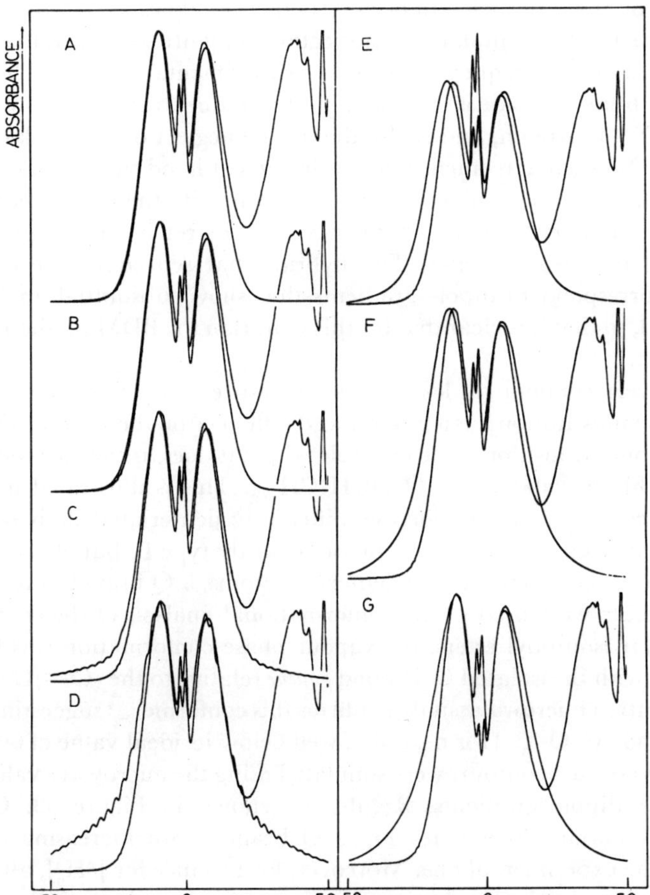

Figure 30 *Contour simulations of the high frequency* $\nu(N{=}O)$ *in methyl nitrite using dipole gradients at different angles* α *with the N=O bond. In the left-hand column the experimental contour is superposed with* trans *hybrids, in the right-hand column* cis *hybrids were used.* (A) $\alpha = 10°$; (B) $\alpha = 20°$; (C) $\alpha = 30°$; (D) $\alpha = 50°$; (E) $\alpha = 40°$; (F) $\alpha = 55°$; (G) $\alpha = 70°$

Published with permission from B. J. van der Veken, R. Maas, G. A. Guirgis, H. D. Stidham, T. G. Sheehan, and J. R. Durig, *J. Phys. Chem.*, 1990, **94**, 4029.

superposing two identical hybrids, with adjusted relative intensity, which were slightly displaced relative to each other. Although the experimental contour is slightly disturbed by the overlap from the low frequency N=O stretch, it is clear that the *trans* hybrids A and B reproduce very satisfactorily the experimental contour. In contrast, no such fit could be obtained with the *cis* hybrids. Hence, the high frequency N=O stretch must be due to the *trans* conformer. Analogous simulations are shown in Figure 31 for the low frequency N=O stretch, at 1620.3 cm^{-1}. Also for this stretch the calculated contours were constructed from two hybrids, in order to reproduce the Q branch structure. Some more excited states are present on the R branch, but because of their weakness no attempts were made to include them in the simulations. Inspection of Figure 31 reveals that simulation *E* reproduces the low frequency part of the experimental contour. The agreement on the high frequency side is less good, but this is due to the overlap with the high frequency N=O stretch, and is also due to the neglect of the weak excited state transitions. No such agreement is found in the left hand side of the figure, where *trans* simulations are shown. Therefore, consistent with the assignment of the high frequency component, the low frequency N=O stretch unambiguously can be assigned to the *cis* conformer. The hybrids that reproduce the experimental contours correspond to dipole gradients that show substantial angles with the N=O bond, so that it is clear that for this vibration the BDM predictions are very poor indeed.

It is not very common, at least when the nature of the vibration is not altered, that isotopic substitution has a pronounced influence on the vapour phase contour of a vibrational transition. One example where it does, is the carbonyl stretch in isopropyl chloroformate, $ClC(O)OCH(CH_3)_2$. This is illustrated in Figure 32, where the contour is shown for five differently deuteriated derivatives. In the normal compound, the contour is predominantly type B, but as more and more hydrogen atoms are replaced by deuterium atoms, a Q branch starts to emerge. These contours were used in the conformational analysis of the compounds.[23]

For similar isopropyl esters, the vapour phase conformation was found to be s-*cis,gauche*, with the isolated C–H bond *gauche* relative to the (O=)C–C(C) bond. A low resolution microwave study confirms this conformer,[24] suggesting that τ, the dihedral angle C–O–C–H is near 40°, well below its ideal value of 60°. Initially, the carbonyl stretch contours were simulated using the microwave value for τ, and using BDM dipole gradients. Results are shown in Figure 33. Clearly, the calculated contours do not develop a Q branch with increasing deuteriation similar to the experimental one. Moreover, for instance for [$^2H_0$], although both experimental and calculated contours are predominantly type B, the PQ,QR separation of the calculated hybrid exceeds the experimental value. Consequently, the simulations in Figure 33 are unsatisfactory. It will be shown below that for the carbonyl stretch in esters the BDM dipole gradient is not very accurate, the actual gradient lying in the ester plane, but showing a small angle to the C=O bond direction.[25] In view of this, it was attempted to improve the above simulations by

[23] B. J. van der Veken and H. H. Liefooghe, *J. Mol. Struct.*, 1994, **328**, 211.
[24] N. S. True and R. K. Bohn, *J. Mol. Struct.*, 1977, **36**, 173.
[25] B. J. van der Veken and H. H. Liefooghe, *J. Mol. Struct.*, 1991, **263**, 177.

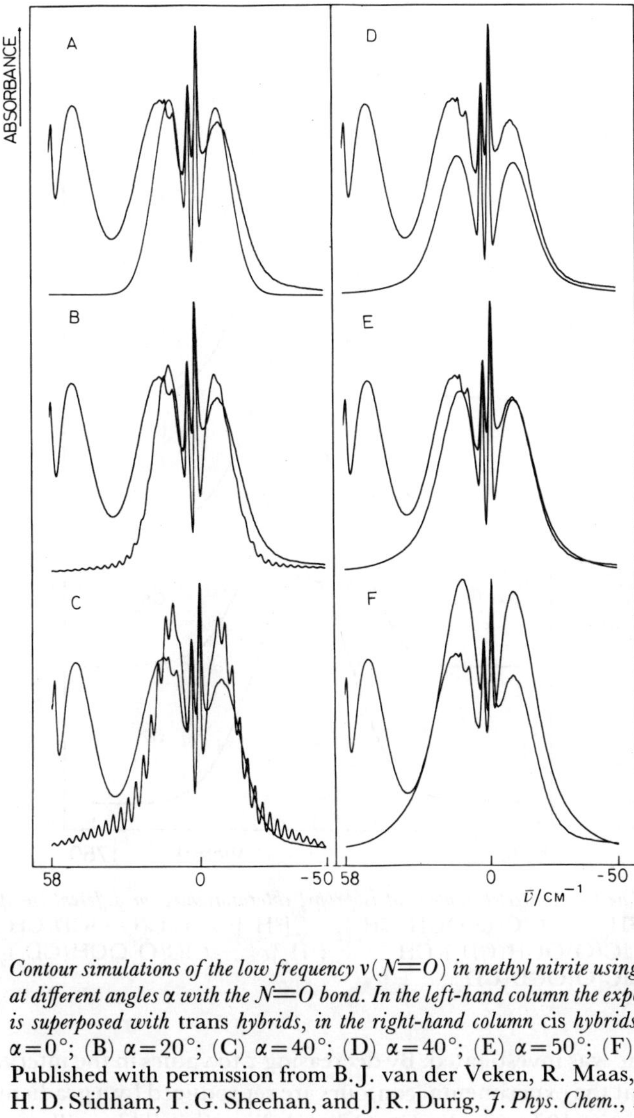

Figure 31 *Contour simulations of the low frequency ν(N=O) in methyl nitrite using dipole gradients at different angles α with the N=O bond. In the left-hand column the experimental contour is superposed with* trans *hybrids, in the right-hand column* cis *hybrids were used.* (A) α=0°; (B) α=20°; (C) α=40°; (D) α=40°; (E) α=50°; (F) α=60°
Published with permission from B. J. van der Veken, R. Maas, G. A. Guirgis, H. D. Stidham, T. G. Sheehan, and J. R. Durig, *J. Phys. Chem.*, 1990, **94**, 4029.

relaxing the dipole gradient. Because of the behaviour of the gradient in other esters,[24] and because there is no obvious coupling by which this gradient would deviate from the ester plane, it was restricted to the plane of the ester function. For none of the possible orientations of the dipole gradient in the ester plane the agreement improves: rotation of the gradient away from the C=O bond in either direction increases the type A content of the hybrid, and because of the higher PR separation of the type A band, the PR separation of the hybrids increases with the type A content, deteriorating the agreement with experiment. Therefore, it had to be concluded that the poor simulations of Figure 33 are due to be the value of τ used to set up the structure of the molecule. Subsequently, the influence of this

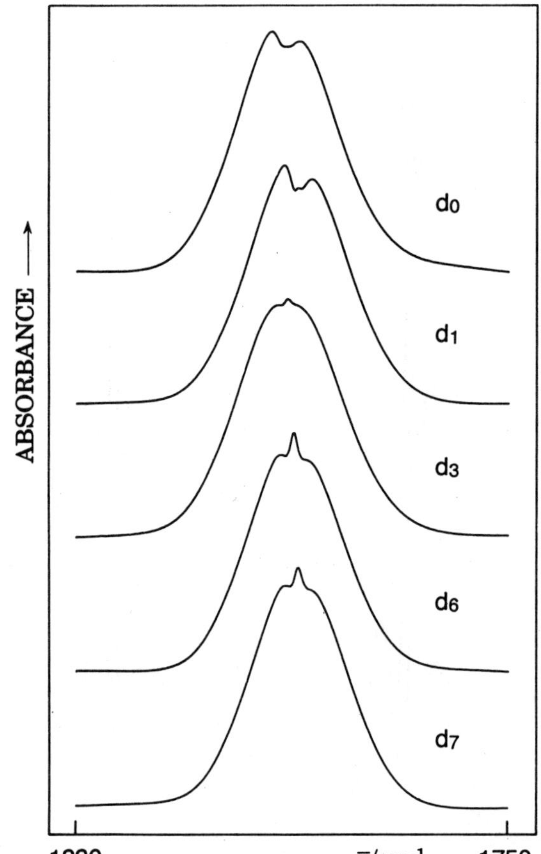

Figure 32 *The C═O stretch contour of isopropyl chloroformate, for different isotopic derivatives.* [$^2H_0$]: $ClC(O)OCH(CH_3)_2$; [$^2H_1$]: $ClC(O)OCD(CH_3)_2$; [$^2H_3$]: $ClC(O)OCH(CD_3)(CH_3$; [$^2H_6$]: $ClC(O)OCH(CD_3)_2$; [$^2H_7$]: $ClC(O)OCD(CD_3)_2$

dihedral angle was investigated. By decreasing τ to values in the interval 20–15° it was found that the experimental contours are reproduced by a gradient that shows an angle of some 10° with the C═O bond direction. This is illustrated for the normal and for the [$^2H_1$] derivatives in Figures 34 and 35. In these figures, hybrids for τ equal to 20° and for varying orientations of the dipole gradient are compared with the experimental contour. It can be seen that for both cases an excellent agreement is obtained for a gradient that shows an angle of 5–10° with the C═O bond. Also, for the other isotopic derivatives equally good simulations were obtained.[23] Therefore, it was concluded that the simulations suggest that in isopropyl chloroformate the isopropyl grouping shows a *quasi-gauche* orientation in which τ, the C–O–C–H dihedral angle, is close to 20°.

A similar pattern emerges for isopropyl cyanoformate, $NCC(O)OCH(CH_3)_2$. Also for this compound a pronounced evolution in the carbonyl stretching contour

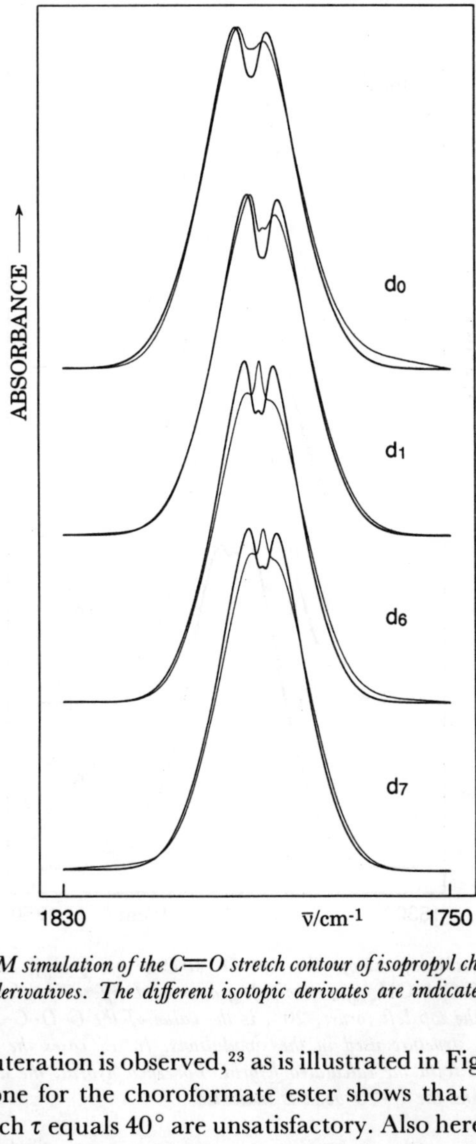

Figure 33 *The BDM simulation of the C═O stretch contour of isopropyl chloroformate, for different isotopic derivatives. The different isotopic derivates are indicated as in Fig. 32*

with increasing deuteration is observed,[23] as is illustrated in Figure 36. An analysis analogous to the one for the choroformate ester shows that the simulations for a conformer in which τ equals 40° are unsatisfactory. Also here the PR separation of the calculated hybrids are too high, and could be improved only by decreasing the value of τ. This is shown in Figure 37, where BDM hybrids for $NCC(O)OCH(CD_3)_2$, for varying angles of τ, are compared with the experimental contour. At each value of τ, reproduction of the experimental contour was attempted by sweeping the dipole gradient across the ester plane. The best agreement was obtained for τ equal to 20°. This, again for the [$^2H_6$] derivative, is illustrated in Figure 38; when the gradient is rotated 20° towards the C–O bond, a very good agreement is obtained. Thus also for isopropyl cyanoformate the simulations suggest that the isopropyl grouping has a *quasi-gauche* orientation, with a value of τ near 20°.

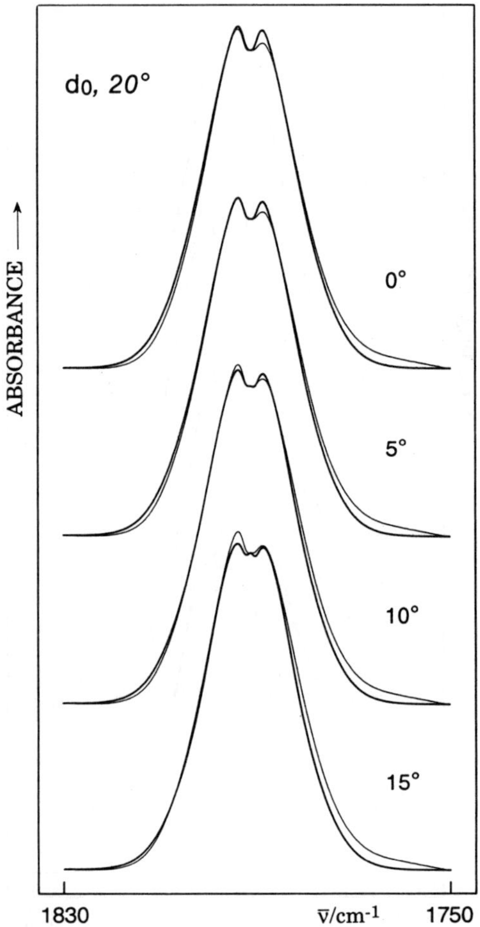

Figure 34 *Contour simulation of the C═O stretch of* [$^2H_0$] *isopropyl chloroformate,* ClC(O)OCH(CH_3)$_2$, *with dipole gradients at different angles to the C═O bond. The angle in the top left corner,* 20°, *is the value of the C–O–C–H dihedral angle of the molecular structure used in the simulations. In all cases the experimental contour is superposed with the calculated hybrid. For each hybrid, the angle between the dipole gradient and the C═O bond is indicated*

11 ANHARMONIC SIMULATIONS

In all previous examples the harmonic approximation was applied, *i.e.* it was assumed that the rotational constants of the upper vibrational state are identical to those of the lower vibrational state. Even in this approximation, a P,Q,R structure is not symmetric: the P branch contains $J+1 \rightarrow J$ transitions, and the R branch contains $J \rightarrow J+1$ transitions. Thus, at the same distance from the band centre, the P branch contains transitions that originate from a higher, and thus less populated, rotational state than the R branch. As a consequence, the R branch is

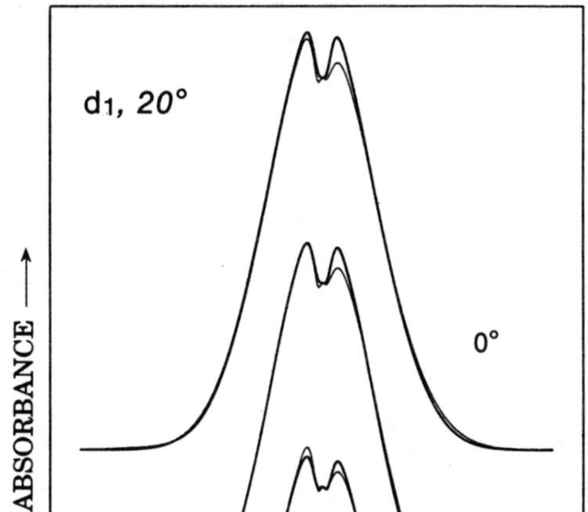

Figure 35 *Contour simulation of the C=O stretch of* $[^2H_1]$ *isopropyl chloroformate,* $ClC(O)OCD(CH_3)_2$, *with dipole gradients at different angles with the C=O bond. The 20° angle is the C–O–C–H dihedral angle of the molecular structure used in the simulations. In all cases the experimental contour is superposed with the calculated hybrid. For each hybrid, the angle between the dipole gradient and the C=O bond is indicated*

somewhat more intense than the P branch. In experimental bands the asymmetry, however, is always bigger than calculated in the harmonic approximation. This is caused by the, small, differences between the rotational constants for the upper and lower vibrational states. These differences, as was already said above, in part would occur even if the molecular vibrations were perfectly harmonic, but it is believed that the larger part of the differences is due to the anharmonicity of the molecular vibrations.[26]

[26] I. M. Mills, in 'Molecular Spectroscopy, Modern Research', ed. K. N. Rao and C. W. Mathews, Academic Press, New York, 1972.

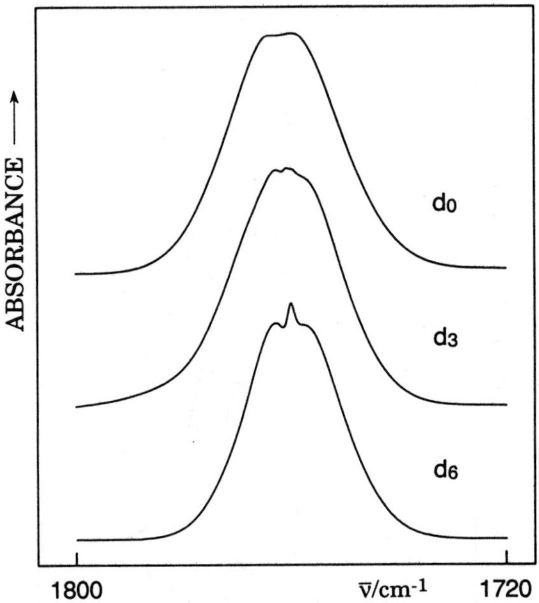

Figure 36 *The C=O stretch contour of isopropyl cyanoformate, for different isotopic derivatives.* [$^2H_0$]: NCC(O)OCH(CH_3)$_2$; [$^2H_3$]: NCC(O)OCH(CD_3)(CH_3); [$^2H_6$]: NCC(O)OCH(CD_3)$_2$

One way of expressing the vibrational dependence of the rotational constants is the following:[26]

$$A_v = A_e - \sum_r \alpha_r^A \left(v_r + \frac{d_r}{2} \right) + \ldots \quad (36)$$

$$B_v = B_e - \sum_r \alpha_r^B \left(v_r + \frac{d_r}{2} \right) + \ldots \quad (37)$$

$$C_v = C_e - \sum_r \alpha_r^C \left(v_r + \frac{d_r}{2} \right) + \ldots \quad (38)$$

In these expressions, A_v, B_v and C_v are the rotational constants for the vibrational state characterized by the vibrational quantum numbers v_r (r ranging from $1\text{–}3n-6$), A_e, B_e and C_e are the rotational constants for the equilibrium structure of the molecule, d_r is the degeneracy of vibrational mode r, and the α_r^X are constants. In first approximation, the latter can be expressed as a sum of three terms:[27] a harmonic term, due to the harmonic part of the vibrational potential, a Coriolis term (to which we here will pay no further attention because for simulation we select bands free from Coriolis coupling), and an anharmonic term containing the cubic force constants of the molecules. Using this expression for α_r, when the force field of the molecule is known through the cubic terms, the α_r, and

[27] H. W. Kroto, 'Molecular Rotation Spectra', J. Wiley and Sons, London, 1975.

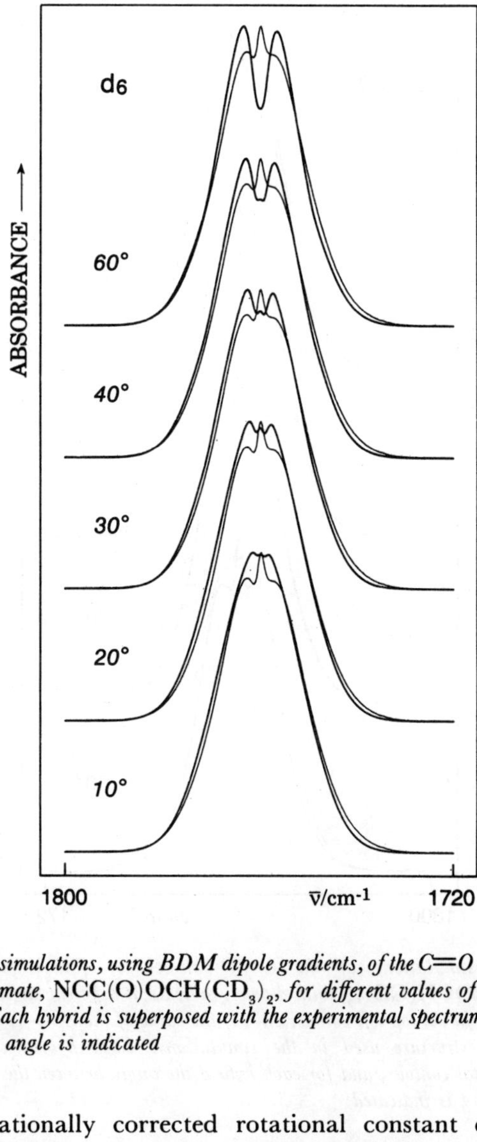

Figure 37 *Contour simulations, using BDM dipole gradients, of the C=O stretch of* $[^2H_6]$ *isopropyl cyanoformate,* $NCC(O)OCH(CD_3)_2$*, for different values of the C–O–C–H dihedral angle. Each hybrid is superposed with the experimental spectrum, and for each hybrid the dihedral angle is indicated*

from it the vibrationally corrected rotational constant can be calculated. Unfortunately, cubic force constants are rarely available. They can be calculated from an advanced *ab initio* calculation, but these calculations are rather involved and not always accessible. Therefore, an approximate way to account for vibrational anharmonicity has been suggested.[20] The approximation is based on the local mode description in terms of Morse oscillators of molecular vibrations.[28] In short, a normal mode is represented by an anharmonic oscillator of the Morse type, for which the potential energy is given by:

$$V(r) = \mathrm{D}\ [1 - \exp\ (-\mathrm{a}\Delta r]^2 \qquad (39)$$

[28] I. M. Mills and A. G. Robiette, *Mol. Phys.*, 1985, **56**, 743.

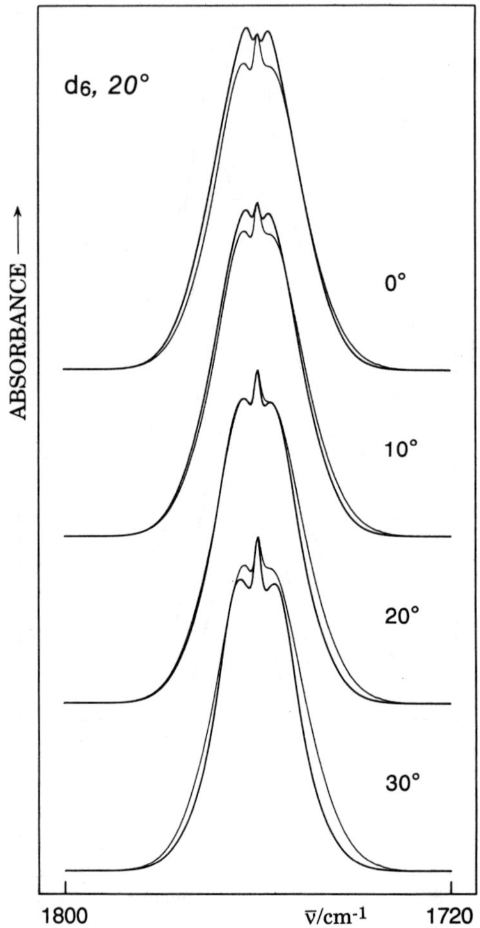

Figure 38 *Contour simulations of the C═O stretch of* [$^2H_6$] *isopropyl cyanoformate,* $NCC(O)OCH(CD_3)_2$, *with dipole gradients at different angles with the C═O bond. The angle in the top left corner,* 20°, *is the value of the C–O–C–H dihedral angle of the molecular structure used in the simulations. Each hybrid is superposed with the experimental contour, and for each hybrid the angle between the dipole gradient and the C═O bond is indicated*

In this expression, Δr is the displacement of the oscillator from its equilibrium, and a and D are constants. The transition frequencies $\nu(\mathrm{v})$ for a Morse oscillator with vibrational quantum number v are given by the following expression:

$$\nu(\mathrm{v}) = (\omega + x)\mathrm{v} + x\mathrm{v}^2 \tag{40}$$

in which:

$$\omega = \hbar\, (2\mathrm{a}^2\mathrm{D}/\mu)^{1/2}/hc \tag{41}$$

and:

$$x = -a^2\hbar^2/2(\mu hc) \quad (42)$$

with μ the reduced mass. The parameters D and a of the Morse function can be calculated if the fundamental and at least one overtone frequency is known. Next, the normal vibration is assumed to be described by a single symmetry coordinate. It is a simple exercise to calculate the rotational constants as function of this symmetry coordinate and thus, in view of the approximation made, the rotational constants are known as numerical functions of the normal coordinate. The wave functions in this normal coordinate for the lower and upper vibrational states of the transition are the corresponding Morse eigenfunctions. Thus, the rotational constants can be numerically averaged over the vibrational wavefunctions (in this model no α_r terms are generated). In the following example we will see how this approximate vibrational averaging improves the simulations.

The Morse–anharmonic model has been investigated in detail using the carbonyl stretching contours in some simple esters,[25] and we will use methyl formate as an example. The simulations, in the harmonic approximation, of this vibration are shown in Figure 39. Clearly, the dipole gradient orientation predicted by the BDM is not fully satisfactory: the agreement improves by rotating the gradient 20° towards the formate C–H bond. For the simulation it can be seen that the experimental R branch is more intense than the calculated one, while the experimental P branch is less intense than the calculated one; such differences are typical for harmonic simulations. For this vibration, the Morse function parameters were least-squares adjusted from the frequencies of the fundamental and of the first and second overtones. The stretching mode was approximated by the C=O stretching internal coordinate, and using this, the rotational constants were calculated as function of the carbonyl stretching mode. Then the rotational constants were numerically averaged over the Morse wavefunctions for the ground and the first, second and third excited vibrational states. Using the rotational constants of the ground and the first excited state, the pure type transitions were calculated, and the fundamental in Figure 40 was constructed, using the same dipole gradient orientation as for the harmonic calculation. It is clear that with the anharmonic contour, a virtually quantitative reproduction of the experimental one is obtained. The anharmonic influences on rotational constants increase with the vibrational quantum number, and the quality of the model is, therefore, better judged from the simulation of overtones. The intensity of overtones is determined not only by the dipole gradient $(\partial\vec{\mu}/\partial Q_i)_0$, but also by the higher order derivatives of the dipole moment, such as $(\partial^2\vec{\mu}/\partial Q_i\partial Q_j)_0$, *etc.* It can be shown, however, that the hybrid for an overtone can still be written as a sum of contributions of the type A, B and C spectra, albeit that the interpretation of the coefficients as the squares of the direction cosines of the dipole gradient is no longer valid. In Figures 41 and 42, hybrids with varying composition for the first and second overtone of the carbonyl stretching are superposed with their respective experimental contour. As expected, the experimental contours show an increasing degree of asymmetry. In these figures, the hybrids are characterized by

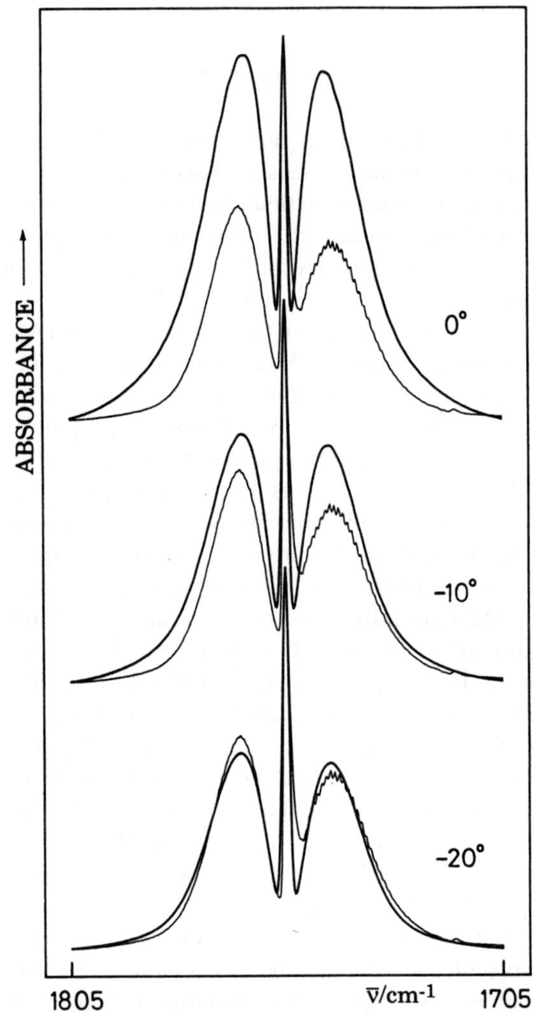

Figure 39 *Harmonic contour simulations of ν(C=O) in methyl formate using s-*cis *dipole gradients at different angles with the C=O bond. Each hybrid is superposed with the experimental contour, and for each hybrid the angle between the gradient and the C=O bond is specified* Published with permission from B. J. van der Veken and H. H. Liefooghe, *J. Mol. Struct.*, 1991, **263**, 177.

the angle which a vector would have if the composition of the hybrid was interpreted in terms of direction cosines. As said above, this interpretation is physically meaningless, and the angle is merely a convenient way to characterize the hybrid. It can be seen in Figures 41 and 42 that for both the first and the second overtone, a hybrid can be constructed that accurately reproduces the experimental contour. This demonstrates that, at least for carbonyl stretches, the anharmonic model based on the Morse potential can be used to improve the quality of a simulation.

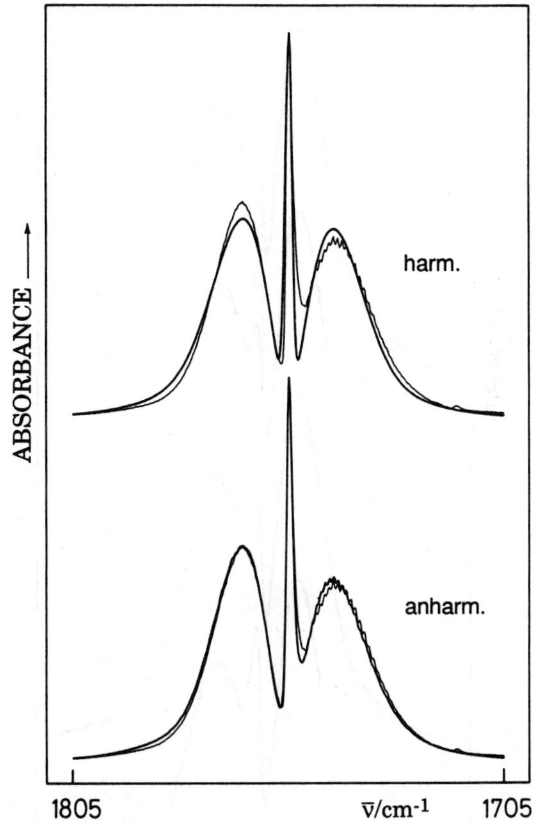

Figure 40 *Harmonic and anharmonic simulations of the C=O stretch of methyl formate, using* s-cis *hybrids for a dipole gradient at* 20° *from the C=O bond. Each hybrid is superposed with the experimental contour*
Published with permission from B. J. van der Veken and H. H. Liefooghe, *J. Mol. Struct.*, 1991, **263**, 177.

12 DIPOLE GRADIENT ORIENTATIONS FOR ISOLATED C–H STRETCHES

It has been discussed above that for vibrations for which the bond dipole model (BDM) does not predict an acceptable dipole gradient, the experimental contour can be reproduced by varying the relative contributions of the type vibrations. This amounts to deducing the coefficients in the expression (32) for a hybrid contour:

$$H(\bar{\nu}) = \cos^2\alpha\ \mathrm{A}(\bar{\nu}) + \cos^2\beta\ \mathrm{B}(\bar{\nu}) + \cos^2\gamma\ \mathrm{C}(\bar{\nu}) \qquad (32)$$

Thus, up to their sign, from such an optimization the direction cosines of the dipole gradient in the principal axis system are obtained from the experimental contour.

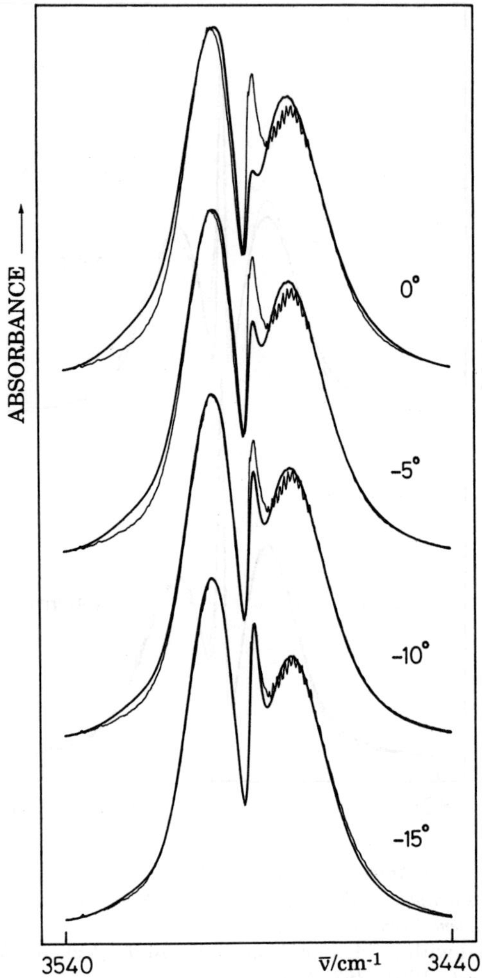

Figure 41 *Anharmonic contour simulations of the first overtone of ν(C=O) of methyl formate, using hybrids of different composition. The angles indicated specify the composition of the hybrids (see text). Each hybrid is superposed with the experimental contour*
Published with permission from B. J. van der Veken and H. H. Liefooghe, *J. Mol. Struct.*, 1991, **263**, 177.

There are not many methods by which this kind of information can be extracted from experiment: the only other way is from an analysis of the electro-optical parameters.[29,30] The latter are the derivatives of the dipole components, in an arbitrary axis system, with respect to internal coordinates. These parameters are calculated from integrated IR intensities, and, because the latter depend on the square of the dipole gradients, their calculation requires intelligent (very often based on *ab initio* calculations) guesses of a substantial number of signs. When a full

[29] W. B. Person and J.H. Newton, *J. Chem. Phys.*, 1974, **61**, 1040.
[30] J. H. Newton and W. B. Person, *J. Chem. Phys.*, 1976, **64**, 3036.

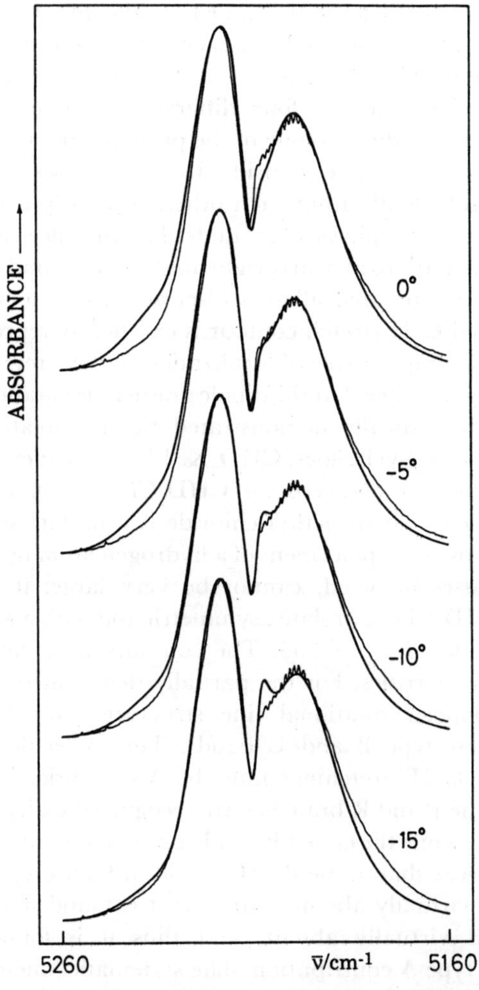

Figure 42 *Anharmonic contour simulations of the second overtone of ν(C═O) of methyl formate, using hybrids of different composition. The angles indicated specify the composition of the hybrids (see text). Each hybrid is superposed with the experimental contour*
Published with permission from B. J. van der Veken and H. H. Liefooghe, *J. Mol. Struct.*, 1991, **263**, 177.

set of electro-optical parameters of a molecule is known, the components, and hence the orientation, of the dipole gradient $(\partial\vec{\mu}/\partial Q_i)_0$ for a normal vibration i can be reconstructed.

In the simulation procedure the integrated intensity is not a parameter, and, consequently, the information obtained on dipole gradient orientations is completely independent from that obtained from electro-optical parameters.

Unfortunately, the squares of the direction cosines in the above equation prevent the orientation of a dipole gradient being determined unambiguously: in the general case three signs are lost in the squaring of the cosines, so a total of $2^3=8$

different vectors $(\partial\vec{\mu}/\partial Q_i)_0$ lead to the same hybrid contour. Not all these vectors have different orientations, although for each orientation the gradient can have two opposite directions. Therefore, as long as the sign of $(\partial\vec{\mu}/\partial Q_i)_0$ is not required, a successful simulation leads to four different, possible orientations. When symmetry restricts the gradient to one of the principal planes, only two different orientations produce the same hybrid. In those cases where the possible orientations differ sufficiently from each other, it usually is possible to reject a number of orientations on physical grounds. For instance, for a well-localized stretching mode, orientations at nearly right angles to the bond are rather unlikely.

For the partially deuteriated alkanes discussed above it was found that the experimental isolated C–H stretch contour is excellently reproduced with BDM dipole gradients. The high degree of localization of the stretching of this type of bond is no ultimate guarantee that the dipole gradient is parallel to the C–H bond, however. This is dramatically demonstrated by the isolated C–H stretch in partially deuteriated methyl halides, CHD_2X.[31,32] As a first example, in Figure 43 the experimental contour observed for CHD_2Cl is compared with the BDM hybrid. The fully deuteriated methyl chloride is a prolate symmetric top. The disturbance caused by the replacement of a hydrogen atom by a deuterium atom, in view of the masses involved, cannot be very large; it can be, correctly, anticipated that CHD_2Cl is a prolate asymmetric top with a κ value *ca.* -1. The value actually calculated is -0.993. The contours will, therefore, show near-symmetric top characteristics. For the partially deuteriated methyl halides this means that pronounced rotational fine structure must be present on the pseudo-perpendicular type B and C bands. The molecules CHD_2X have C_s symmetry, and the C–H stretchings must be A/C hybrids. Therefore, the fine structure bands in the P and R branches are recognized easily as being due to the type C contribution, while the closed P and R branch intensity is due to the type A contribution. It is clear that in the BDM hybrid in Figure 43 the closed P and R branch intensity is virtually absent, and thus it is found that the BDM and R branch intensity is virtually absent, and thus it is found that the BDM underestimates the type A contribution. The systematic, one-dimensional search of all possible hybrids has been made,[31] and it was found, as is shown in Figure 44, that the experimental contour can be reproduced. The hybrid shown corresponds to the dipole gradients for which ω, the angle of the gradient with the C–H bond, is either 38 ± 20 or $71 \pm 2°$. The latter value was judged unrealistically high, and the value of $38 \pm 2°$, with the gradient rotated towards the bisector of the methyl group, was adopted. Corroborating evidence was obtained from electro-optical parameter calculations: using the literature data for the latter, the angle between the gradient and the C–H bond was calculated to be 37°. This shows that the two methods lead to consistent results.

Similarly, the contours of the other halides were analysed.[31,32] As an illustration, hybrids for different orientations of the dipole gradient for CHD_2F are compared with the experimental contour in Figure 45, with the hybrids characterized by the value of ω for which they were calculated. Inspection of the intensities of the fine

[31] A. Y. Grupping, D. C. McKean and B. J. van der Veken, *J. Mol. Struct.*, 1993, **295**, 273.
[32] A. Y. Grupping, D. C. McKean and B. J. van der Veken, *J. Mol. Struct.*, 1993, **300**, 313.

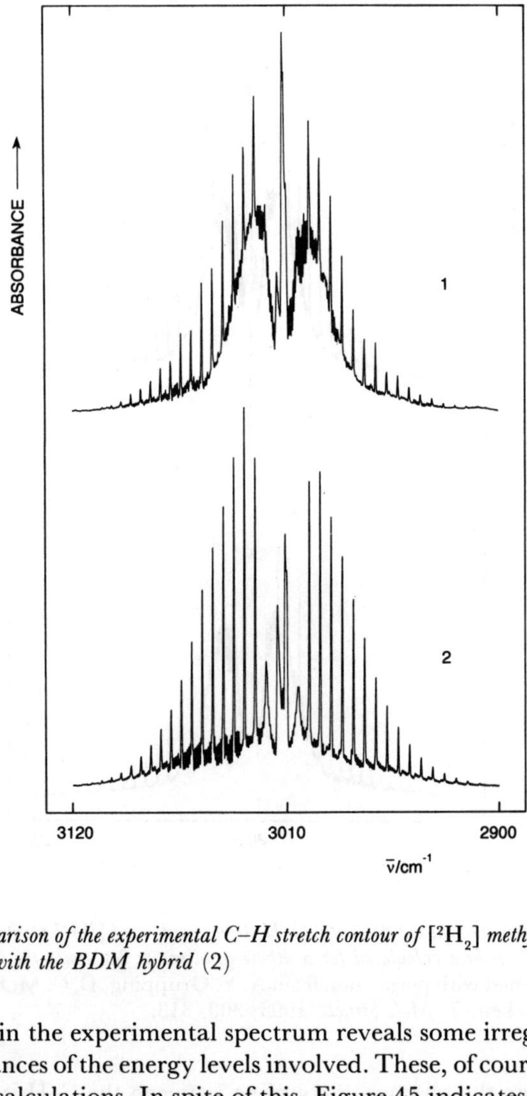

Figure 43 *Comparison of the experimental C–H stretch contour of* [$^2H_2$] *methyl chloride,* CHD_2Cl (1), *with the BDM hybrid* (2)

structure bands in the experimental spectrum reveals some irregularities, caused by local disturbances of the energy levels involved. These, of course, have not been included in the calculations. In spite of this, Figure 45 indicates that the angle ω must lie between 10 and 20° – the value of $15 \pm 5°$ was eventually adopted. Also for CHD_2F the result was confirmed, in this case from *ab initio* calculations,[33] from which a value for ω of 15° can be deduced.

The significant difference between the ω values for CHD_2F and CHD_2Cl suggests that the nature of the halogen is the determining factor. In line with this, the value of ω was found to be even higher in the bromide, $\omega = 44 \pm 2°$, and in the iodide, $\omega = 49 \pm 3°$. The reorienting power of the halide must be related to the polarizability of its electron cloud. As the C–H bond stretches, the electrical phenomena accompanying it induce a rearrangement of the electron density in the

[33] T. K. Ha, D. Luckhaus and M. Quack, *Chem. Phys. Lett.*, 1992, **190**, 590.

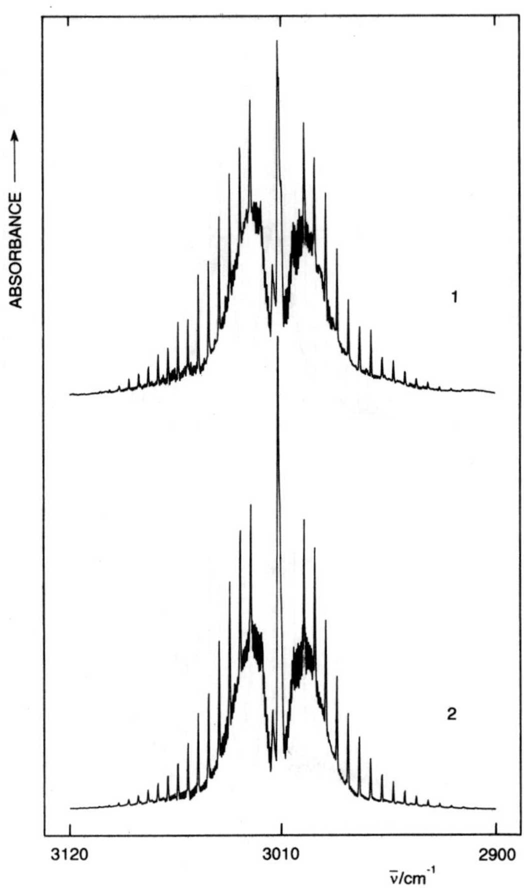

Figure 44 *Comparison of the experimental C–H stretch contour of* [$^2H_2$] *methyl chloride*, CHD_2Cl, *with the hybrid calculated for a dipole gradient at* 38° *with the C–H bond*
Published with permission from A. Y. Grupping, D. C. McKean, and B. J. van der Veken, *J. Mol. Struct.*, 1993, **300**, 313.

halide atom. From the absence of coupling between the C–H stretching and the carbon–halide stretching, it follows that this rearrangement must take place predominantly in the free electron pairs of the valence shell of the halide electron.

The important conclusion that, in the present context, must be drawn from this is that even for highly characteristic modes such as isolated C–H stretches the dipole gradient can substantially deviate from BDM predictions. From the comparison of the alkanes with the methyl halides it follows that deviations arise when electron rich atoms are bonded to the carbon atom of the isolated C–H bond, and that the magnitude of the deviation is determined by the polarizability of the substituent.

This conclusion was tested against some more examples.[14] A problem similar to the methyl halides is encountered in partially deuteriated propyl bromides. In the

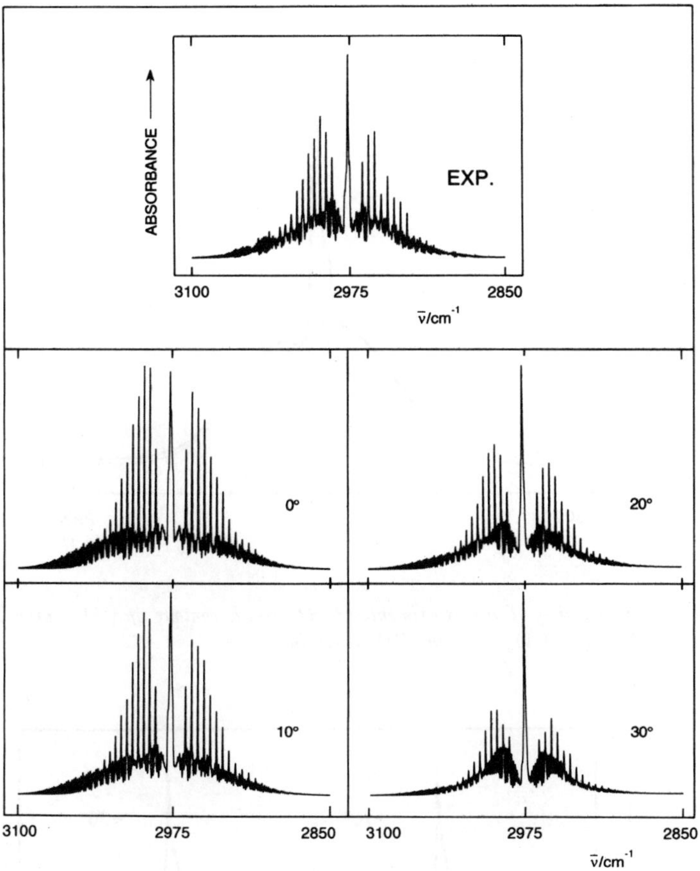

Figure 45 *Comparison of the experimental C–H stretch contour of* [$^2H_2$] *methyl fluoride*, CHD_2F *(EXP), with hybrids calculated for dipole gradients at different angles with the C–H bond. For each hybrid, the angle between the gradient and the C–H bond is indicated*

isomer $CD_3CHBrCD_3$ only one conformer is possible, and the C–H bond can have only one orientation relative to the rest of the molecule. Consequently, a single C–H stretch is expected. The contour of this transition is shown, superposed with the BDM hybrid, in Figure 46. As can be seen, with the BDM dipole gradient a reasonable, but not fully satisfactory agreement is obtained. This compound has C_s symmetry, with the C–H bond lying in the (a,c)-plane, and thus the C–H stretch must have an A/C hybrid contour. In Figure 47 hybrids showing increasing angles between the dipole gradient and the C–H bond are compared with the experimental contour. It is obvious that the best simulation is obtained for a gradient that has been rotated 20° away from the C–H bond (the C–H bond in this molecule is almost parallel to the *c*-axis, so the direction in which the gradient is rotated is unimportant). For this derivative, the bromine atom is bonded to the carbon atom of the C–H bond, and again it is found that the dipole gradient for the

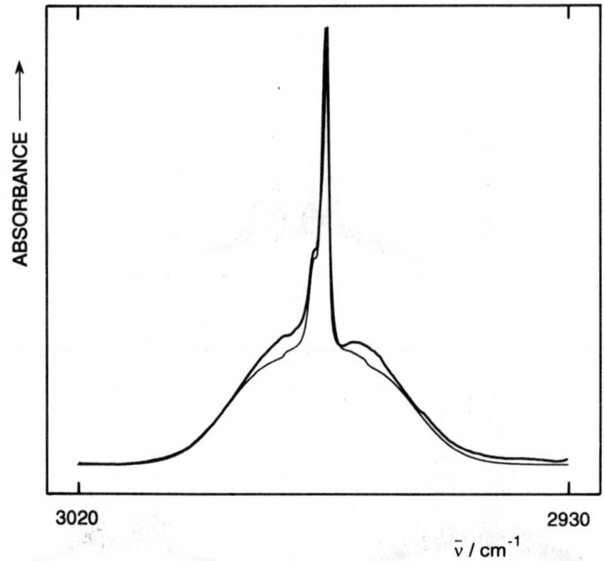

Figure 46 *Superposition of the experimental C–H stretch contour of* $[^2H_6]$ *prop-2-yl bromide,* $CD_3CHBrCD_3$*, with the BDM hybrid*

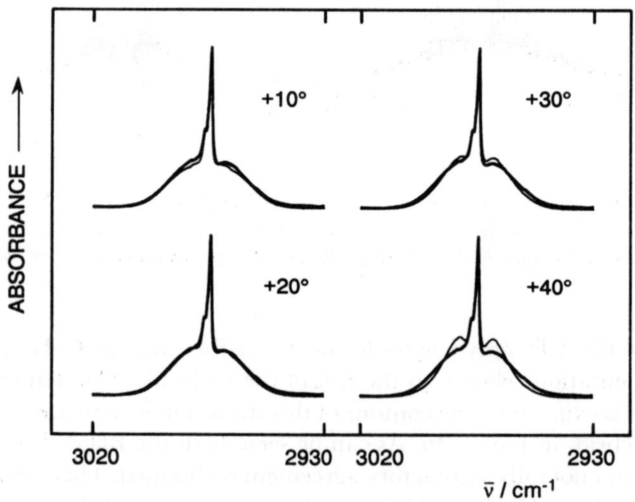

Figure 47 *Harmonic contour simulations of the C–H stretch contour of* $[^2H_6]$ *prop-2-yl bromide,* $CD_3CHBrCD_3$*, using dipole gradients at different angles with the C–H bond. Each hybrid is superposed with the experimental contour, and the angle between the dipole gradient and the C–H bond is indicated.*

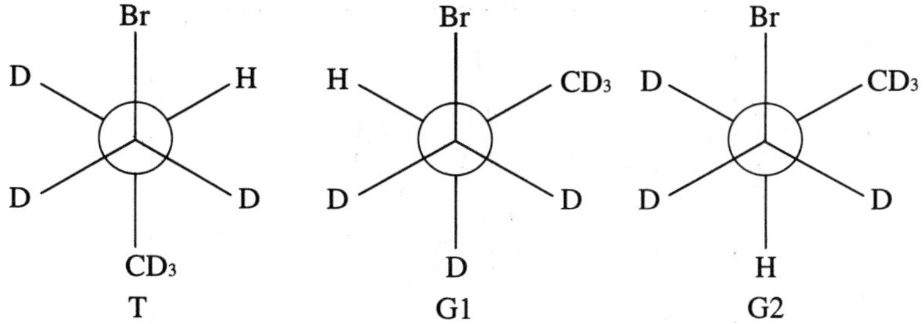

Figure 48 *Newman projections of the different orientations of the C–H bond in* [$^2H_6$] *prop-1-yl bromide*, CD_3CHDCD_2Br

C–H stretch substantially differs from the BDM predictions. The value of ω obtained for this compound is significantly smaller than the one obtained for CHD_2Br: it is supposed that the CD_3 groupings in the propyl bromide shield the bromine atom from the events taking place in the C–H bond during its stretching.

Next, it is interesting to investigate how far the reorienting influence of an electron rich substituent reaches. This can be investigated using another isomer, prop-1-yl bromide. In the isotopic derivative CD_3CHDCD_2Br, the bromine atom is not substituted on the carbon atom of the C–H bond, but is one carbon atom away. This isomer occurs in the vapour phase as a mixture of *trans* and *gauche* conformers. This means that in total three different C–H stretches must be expected, one in the *trans* conformer, and two in the *gauche* conformer. The Newman projections for the different orientations are given in Figure 48. In the experimental C–H stretching contour three transitions can be recognized, and these were simulated using BDM hybrids. The results of this are compared with the experimental contour in Figure 49. From the superposition of the experimental with the calculated sum profile, given in the lower half of Figure 49, it was concluded that the agreement could not be significantly improved by relaxing the dipole gradient orientations. This result suggests that the reorienting power of an electron-rich substituent is a short range phenomenon. This was confirmed by the simulations of, for instance, the C–H stretching contour in CHD_2CD_2Cl,[14] where again a very good agreement is obtained with the BDM dipole gradients.

Although not all factors influencing the orientation of a dipole gradient are understood, at least for isolated C–H stretches a pattern emerges: the BDM prediction of the dipole gradient holds well if no electron-rich substituents are present in the immediate vicinity, *i.e.* bonded to the same carbon atom, while the gradient is poorly predicted when such substituents are attached to the carbon atom.

In the previous examples, the deviations from the BDM were caused by halogen atoms. These are not the only electron rich atoms that can be found in organic molecules. Far more simple molecules contain oxygen, which also contains free pairs, and it has to be expected that isolated C–H bonds adjacent to oxygen atom will also exhibit dipole gradients that are not parallel to the bond. This was indeed found to be the case in partially deuteriated methyl nitrate, CHD_2ONO_2. This compound has a planar heavy atom skeleton,[34] and the hydrogen can take up a

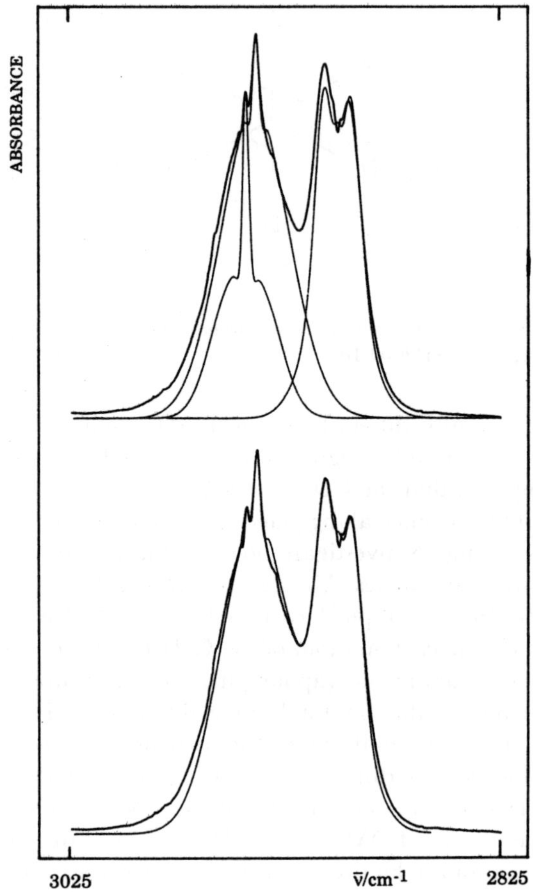

Figure 49 *Experimental C–H stretch contour of* CD_3CHDCD_2Br *superposed with individual BDM hybrids (not drawn to scale)* (1), *and with the calculated sum profile* (2)

position in the heavy atom plane, which we call the in-plane position, and it can also be found in two equivalent out-of-plane positions. These positions are shown in the principal axis projections in Figure 50. Because of overlap of the oxygen free electron pairs with the anti-bonding σ orbitals of the out-of-plane C–H bonds, the latter are weaker than the in-plane C–H bond, so that the out-of-plane C–H stretch is found at lower frequency than the in-plane stretch.[34] The vapour phase infrared spectrum, although disturbed on the low frequency side by overlap with an overtone, shows two well-defined transitions, which are assigned in the order predicted above. Each of the transitions shows a complex Q branch: the high frequency C–H stretch had to be simulated using four hybrids, while the low frequency transition was simulated using two hybrids. The experimental contour is compared with the BDM simulation in Figure 51. The relative intensities of the

[34] B.J. van der Veken, M.A. Herman and D.C. McKean, *J. Mol. Struct.* 1984, **114**, 27.

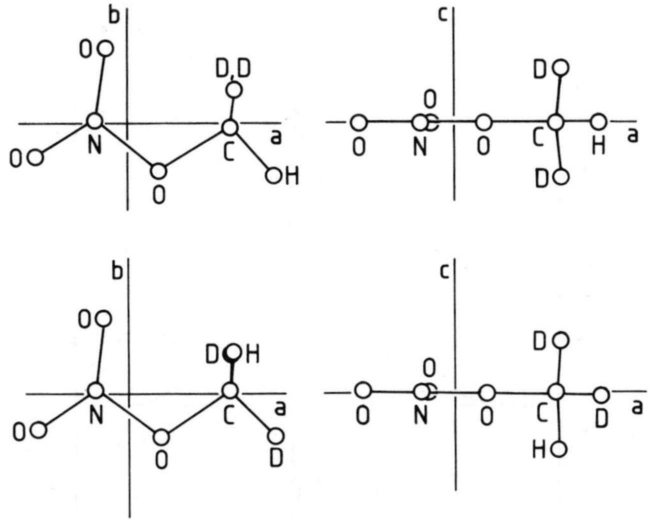

Figure 50 *Principal axis projections of the two different conformers of* [$^{2}H_{2}$] *methyl nitrate,* CHD_2ONO_2

transitions were adjusted so that in the calculated sum contour the Q branch maxima coincide with the experimental ones: it is clear that the calculated contour shows intensity deficiencies in the P and R branches, indicating poor dipole gradient orientations. The derivative with the in-plane C–H bond has C_s symmetry, and all possible hybrids are generated in a one-dimensional scan. The strong overlap of the corresponding experimental contour with the out-of-plane contour, however, prevents the optimization of this contour being performed individually: first the more intense out-of-plane contour must be simulated more accurately. The gradient for the out-of-plane C–H stretch is not restricted by symmetry, and its optimization requires a two-dimensional search. Such a search can be systemized in the following way. By shifting the gradient to the centre of the principal axis system, the end points of all possible gradients describe a sphere. The principal axis planes divide the sphere into eight segments. For a chosen set of absolute values of the direction cosines of the gradient, the all-positive sign combination leads to a gradient in the first segment, delimited by the positive sides of the inertial axes. A gradient in one of the other seven segments is related to one in the first segment by a different sign combination of the direction cosines. Hence, from what has been said above, it follows that all possible contours are generated from dipole gradients in the first segment. Consequently for the simulation of an arbitrary hybrid, only the first segment of the sphere must be scanned. The signs of the principal axis can be chosen such that the out-of-plane C–H bond is contained in the first segment. The orientation of this bond is described by the unit vector $\vec{\mathbf{s}}$. The orientation of a dipole gradient can then be specified using angles ϕ and θ. The former is the angle between the projection in the *a,c*-plane of the gradient and $\vec{\mathbf{s}}$. The value of ϕ is positive in the direction in which, starting from the projection of $\vec{\mathbf{s}}$, cos α of the gradient increases. The angle θ is the difference of the angles θ_1

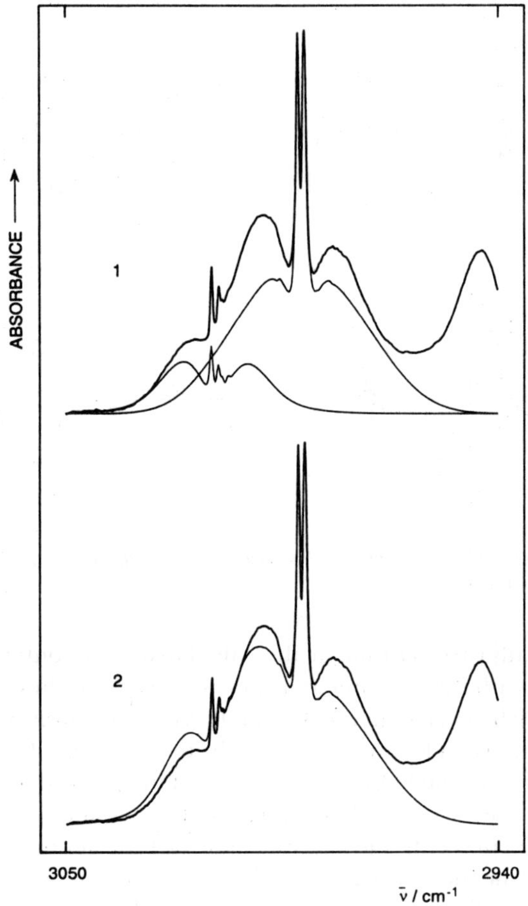

Figure 51 *The C–H stretching region of* [$^2H_2$] *methyl nitrate*, CHD_2ONO_2. (1) *Superposition of experimental contour with individual BDM hybrids (not drawn to scale)*; (2) *superposition of experimental contour with calculated BDM sum contour*

between $\vec{s}$ and the b-axis of the inertial system, and the angle θ_2 between the dipole gradient and the b-axis. The angle θ is positive in the direction in which, starting from $\vec{s}$, $\cos\beta$ of the gradient increases. Using these conventions, ϕ and θ are 0 when the gradient is parallel to the C–H bond. For the C–H stretching in partially deuteriated methyl nitrate, contours were constructed systematically for values of ϕ and θ having an interval of 10°. The cases investigated are indicated on the projection of the segment of the sphere of all possible gradients in Figure 52. A number of the calculated hybrids are compared with the experimental contours in Figures 53 and 54. The simulations of Figure 53 show the variation of the contour with θ, for ϕ equal to 0°, while those of Figure 54 illustrate the variation with ϕ, for θ equal to 10°. For the simulations in this search, the best agreement was found for $\theta = 0°$ and $\phi = 30°$. The neighbourhood of this point was searched with a 5°

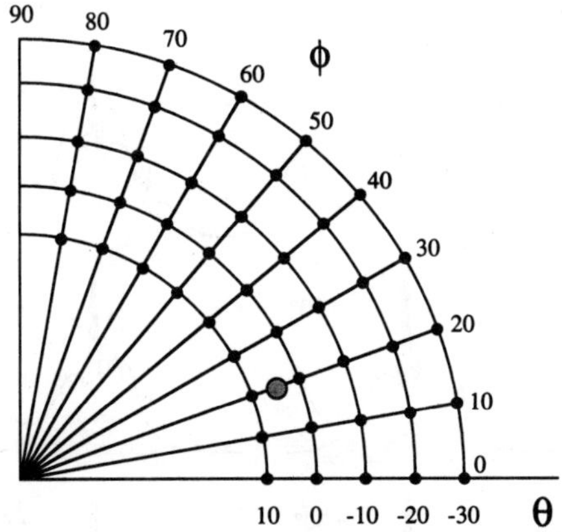

Figure 52 *Summary of the dipole gradient orientations (full dots) used in the simulation of the out-of-plane C–H stretch of* CHD_2ONO_2. *The angles* ϕ *and* θ *are defined in the text. The shaded dot at* ($\phi = 20°$; $\theta = 5°$) *specifies the dipole gradient orientation leading to the best simulation of the experimental contour*

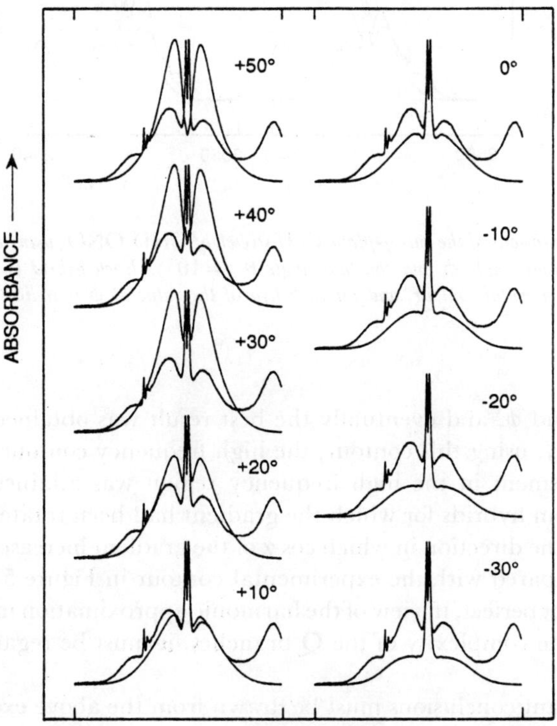

Figure 53 *Simulations of the out-of-plane C–H stretch of* CHD_2ONO_2 *using dipole gradients with variable angle* θ *and constant angle* ϕ (=0°). *Each hybrid is superposed with the experimental contour, and for each hybrid the value of* θ *is indicated*

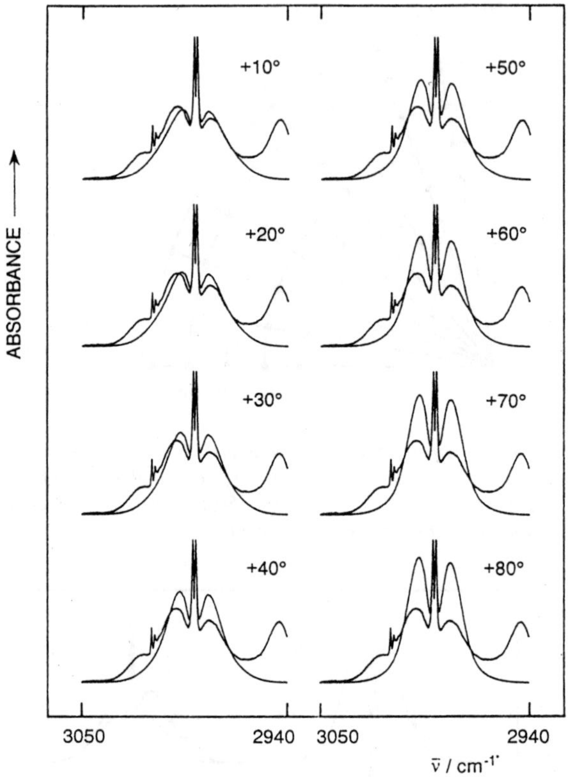

Figure 54 *Simulations of the out-of-plane C–H stretch of* CHD_2ONO_2 *using dipole gradients with variable angle* ϕ *and constant angle* θ $(=10°)$. *Each hybrid is superposed with the experimental contour, and for each hybrid the value of* ϕ *is indicated*

interval of θ and ϕ, and eventually the best result was obtained for $\theta=5°$ and $\phi=20°$. Finally, using this contour, the high frequency contour was optimized. The best agreement in the high frequency region was attained for a contour constructed from hybrids for which the gradient had been rotated 20° out of the C–H bond, in the direction in which cos α of the gradient increases. The final sum contour is compared with the experimental contour in Figure 55. Although the agreement is not perfect, in view of the harmonic approximation made, and taking into account the complexity of the Q branches, it must be regarded as the best possible.

Two important conclusions must be drawn from the above exercise. First, the simulations confirm that for isolated C–H bonds adjacent to an oxygen atom, the dipole gradient for the stretching shows an appreciable angle with the orientation of the bond, in agreement with the expectation that for this type of bond the gradient is poorly predicted by the BDM. Secondly, this application proves that also for two-dimensional searches a satisfactory simulation can be obtained.

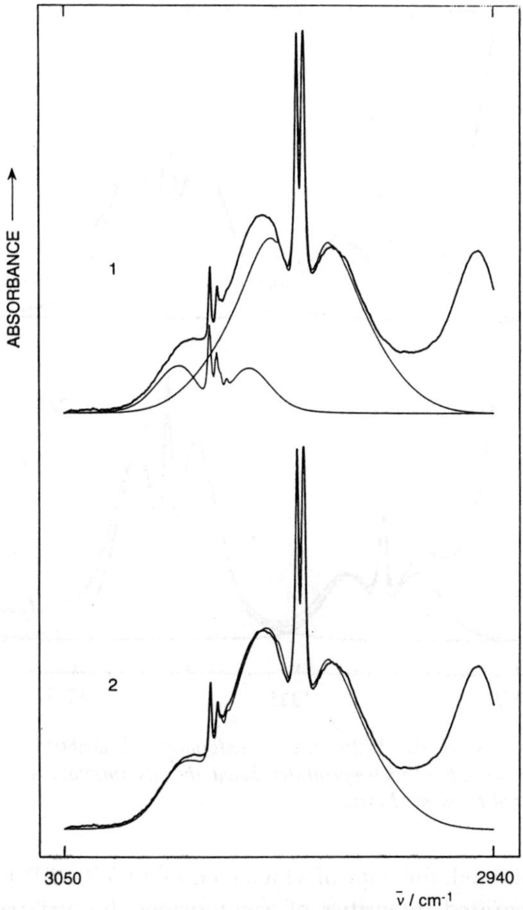

Figure 55 *Comparison of the experimental C–H stretch contour of* CHD_2ONO_2 *with the optimized calculated contour.* (1) *Superposition of experimental contour with individual optimized hybrids;* (2) *superposition of experimental contour with optimized sum contour*

13 DETERMINATION OF INTEGRATED INTENSITIES

In some applications of vapour phase infrared spectroscopy, the integrated intensities of P,Q,R contours are important parameters. When the contour is isolated from others, in general it is easy to define the baseline, and the integration can easily be performed numerically. However, when substantial overlap is present, drawing in a baseline is not a trivial matter, and it becomes very difficult to abstract the integrated intensities. In such cases, contour simulation can come to the rescue.

This can be illustrated with the determination of the vapour phase enthalpy difference between the *gauche* and *trans* conformers of n-butane: when the logarithm of the ratio of the integrated intensity of a *gauche* to that of a *trans* band is plotted against the inverse of the temperature at which the spectrum was run, a

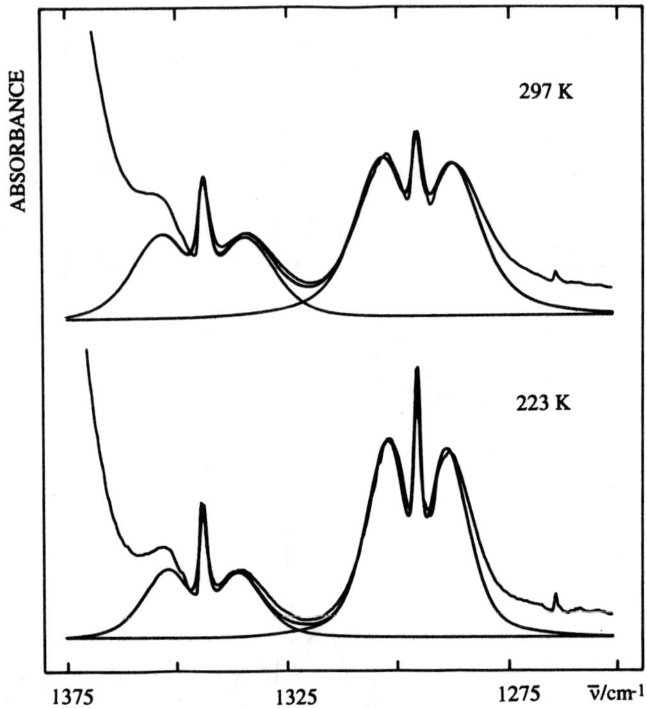

Figure 56 *Simulations of the* 1325 cm^{-1} *conformational doublet of n-butane at different temperatures. For both temperatures shown, the experimental contour is superposed with the optimized harmonic hybrids.*

straight line is obtained, the slope of which is equal to $\Delta H_0/R$.[35] This determination requires the integrated intensities of a conformer doublet recorded at several temperatures. For n-butane, the *gauche* and *trans* conformers have very similar vibrational frequencies, and also within each conformer many modes are near-degenerate. As a consequence, very few well-separated conformer doublets occur in the spectrum. In fact, only the doublet near 1325 cm^{-1} was found suitable for simulation.[36] Although the overlap between the two components of the doublet is not very strong, the situation is complicated on the high frequency side by the presence of an intense complex, which strongly overlaps with the high-frequency component of the doublet. Also, on the low frequency side of the doublet an interfering band is present. In spite of these problems, the 1325 cm^{-1} doublet in the spectra of butane recorded at several temperatures between 297 and 223 K could be simulated. The results at the highest and the lowest temperature are given in Figure 56: in both cases the individual hybrids and the sum contour are compared with the experimental one. It can be seen that in the central region, where the doublet is nearly free from interference, the contour is reproduced

[35] P. J. D. Park, R. A. Pethrick and B. H. Thomas, in 'Internal Rotation in Molecules', ed. W. J. Orville-Thomas, J. Wiley and Sons, London, 1974, p. 57.

[36] W. A. Herrebout, B. J. van der Veken, Aiying Wang and J. R. Durig, *J. Phys. Chem.*, 1995, **99**, 578.

satisfactorily. The relative composition of the doublet having been optimized at each temperature, the integrated intensities were obtained by numerical integration of the individual components. It will be appreciated that abstracting the integrated intensities of the components of the doublet would have been extremely difficult without the contour simulation! Although the experimental contour is reproduced quite well, the success of the simulation depends on the quality of the enthalpy difference calculated from the intensities: for n-butane the ΔH_0 has been determined by a number of other experimental techniques, and the value obtained from the simulations was found to agree very well with the other experimental data.[36]

CHAPTER 8

Industrial Applications of Vibrational Spectroscopy and the Role of the Computer

S. F. PARKER

1 INTRODUCTION

Molecular spectroscopy is widely used in industry. The applications range from quality control and quality assurance of raw materials, to customer complaints, to quantitative analysis and online process monitoring and control. In this chapter the emphasis will be on applications that make use of the power of vibrational spectroscopy to give information, both qualitative and quantitative, about the molecular species present.

In industry, time is invariably at a premium so the pressure is always to carry out the task in the most effective manner possible in order to gain the maximum information from each analysis. Modern instrumentation is almost invariably either run from a computer or is capable of being interfaced to one to transfer data. This means that virtually all analysis, whether it is plotting the spectrum or factor analysis, is computer based. In this chapter, a number of examples have been chosen, from a variety of industries, to illustrate some of the methods that are commonly employed. The underlying theory has been discussed previously in this book, particularly in Chapter 1.

2 EXAMPLES

2.1 Spectral Subtraction

Undoubtedly the most common form of data manipulation in vibrational spectroscopy is spectral subtraction. At its best, the technique is capable of eliminating all of the interfering species, leaving the target species isolated and hence more amenable to identification or further analysis. For the subtraction to be successful, the frequency axis needs to be highly reproducible. The use of a laser referenced interferometer guarantees that this is the case for Fourier transform infrared (FTIR) spectroscopy.

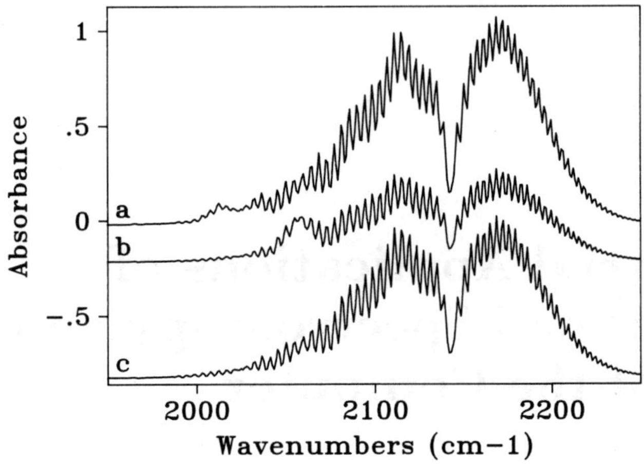

Figure 1 *IR spectra* (256 *scans*, 4 cm^{-1} *resolution*, 50 mbar *pressure*) *of* (a) *a sample taken directly from a CO cylinder*, (b) *syngas sample direct from the cylinder*, (c) *CO after passage through two carbon scrubbers*

2.1.1 The Analysis of Metal Carbonyls in Carbon Monoxide. Nickel tetracarbonyl, $Ni(CO)_4$, and iron pentacarbonyl, $Fe(CO)_5$, are unique in that they are the only metal carbonyls that may be formed from the metals and CO by direct reaction at any significant rate.[1] This means that they are routinely present in high pressure gas cylinders containing CO.

A research project was studying the catalytic conversion of syngas (CO/H_2) to hydrocarbons and there was concern that the metal carbonyls were depositing on the catalyst thus leading to a product that had unusually high levels of metal present. The metal carbonyls are normally removed from the gas stream by passage through an activated carbon scrubber and the problem was to check the efficiency of the filter.

Such an analysis presents a challenging analytical problem because the carbonyls are present at parts per million (ppm) or less. However, metal carbonyls have very strong and characteristic IR spectra and hence FTIR spectroscopy was the chosen method of analysis. Figure 1 shows the IR spectra obtained with a gas cell of 6 m pathlength. Figure 1(*a*) is the spectrum of a sample taken directly from a CO cylinder, (*b*) is the spectrum of a syngas sample direct from the cylinder and (*c*) is the spectrum of CO after passage through two carbon scrubbers.

In Figures 1(*a*) and (*b*) the low frequency side of the envelope of the gas phase CO is distorted indicating the presence of metal carbonyls. Careful inspection of Figure 1(*c*) showed that the spectrum very closely matched the CO spectra in the literature indicating that there was little or no metal carbonyl present. Hence this spectrum was used as a reference CO spectrum to be subtracted from Figures 1(*a*)

[1] F. A. Cotton and G. Wilkinson, 'Advanced Inorganic Chemistry', Wiley Interscience, New York, 3rd edn., 1972, p. 692.

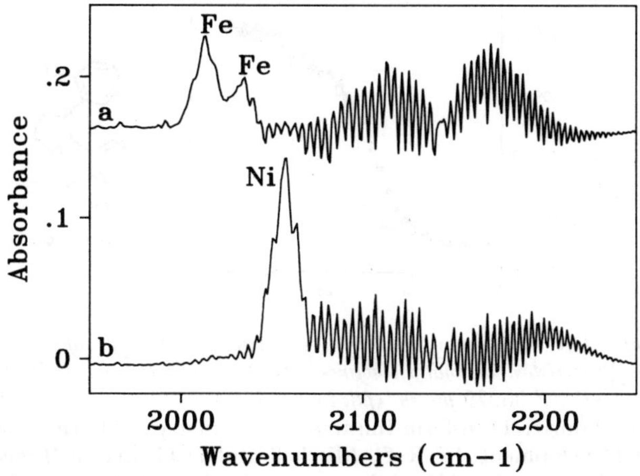

Figure 2 *Difference spectra resulting from the subtraction of the scrubbed CO sample from* (a) *the CO sample direct from the CO cylinder,* (b) *the syngas sample direct from the cylinder*

and (*b*). The results are shown in Figure 2(*a*), the CO sample direct from the CO cylinder, and Figure 2(*b*), the syngas sample direct from the cylinder. The bands at 2013 and 2035 cm^{-1} in Figure 2(*a*), can be assigned to $Fe(CO)_5$ and the band at 2058 cm^{-1} in Figure 2(*b*), can be assigned to $Ni(CO)_4$. It is somewhat surprising that there is predominantly only one type of carbonyl present in each gas; this may well be fortuitous. Quantitation could be carried out using spiked gas samples, but in view of the toxicity of the metal carbonyls, this was not done. Unfortunately, the literature[2,3] values for the extinction coefficients were ambiguous, but indicated that the concentrations were less than 1 ppm. Addition of various amounts of Figures 2(*a*) and (*b*) to the spectra of scrubbed CO and syngas, indicated that there was no more than 10% of the initial carbonyl present and probably much less. This demonstrated that the scrubbers were highly effective and that the source of the metals in the product must lie elsewhere.

This analysis was unusual in that the spectrometer was taken to the sample rather than the other way round. FTIR instruments that are sufficiently small and robust enough to allow them to be taken to the problem, rather than located in a central facility, are now available. This enables greater interaction between the analyst and the people with the problem, resulting in a better understanding of the needs of both parties.

2.2 Digital Data

The ability to have the data available in digital form has many advantages and to an extent all of the forms of manipulation discussed in this chapter rely on this fact. It also leads to an enormous flexibility in the way that the data can be utilized.

[2] R. S. McDowell, *Am. Ind. Hyg. Ass. J.*, 1971, **32**, 621.
[3] A. W. Mantz, *Appl. Spec.*, 1976, **30**, 539.

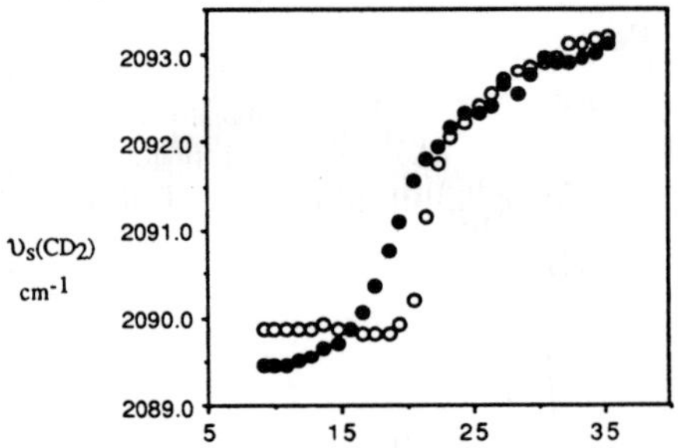

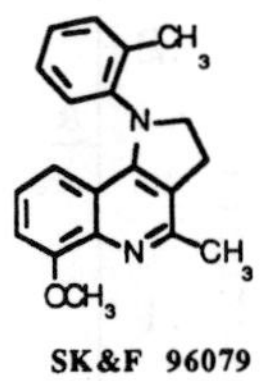

Figure 3 *Temperature dependence of the C–D stretch, $\nu(CD_2)$, of the phosphoglyceride* [$^2H_{54}$]*dimyristoylphosphatidylcholine- in D_2O (open circles), and in D_2O with* 0.125 mol dm^{-3} *SK&F* 96079 *present (filled circles)*
From 'Fourier Transform Infrared Spectrometry in Pharmaceutical Research and Development' by R. C. Mitchell and D. C. Lee in 'Proceedings of the Second International Workshop on Fourier Transform Infrared Spectroscopy', University of Antwerp, 1992, p. 53, reproduced with permission.

2.2.1 In-situ *Studies of Lipids*. One of the most common uses of spectroscopy is to follow the evolution of a reaction as a function of time, temperature or concentration. The use of computer systems is essential in order to cope with the sheer volume of data that is produced. Gas chromatography–mass spectroscopy (GC–MS) and other hyphenated techniques are a case in point. A GC–MS experiment can easily produce 40 megabytes of data in one experiment. To analyse this by hand would make the results irrelevant on a typical industrial time scale (a few days at most and often only an hour or two), as well as cost ineffective.

One of the strengths of IR spectroscopy is the ability to follow processes that involve phase changes, especially those that cause a change of state. Thus it provides a complementary technique to differential scanning calorimetry (DSC). In the pharmaceutical industry it has been used to study model membranes such as liposomes (a reverse micelle consisting of an aqueous core and a phospholipid shell). The high precision of FTIR instruments is essential in these cases because they involve the measurement of small wavenumber shifts and absorbance changes with temperature. Studies of the interactions of a possible drug with model membranes are useful for assessing the ability of the compound to disrupt the membrane and is an indication of possible toxicity problems.

Figure 3[4] shows a plot of the symmetric C–D stretch, $\nu_s(CD_2)$, of an isotopically labelled phosphoglyceride, [$^2H_{54}$] dimyristoylphosphatidylcholine, in D_2O. Below 20 °C, the C–D stretch was at 2090 cm^{-1}, which indicated that the hydrocarbon side-chains in the bilayers were in the gel state and contained a high proportion of *trans* conformers. Between 20 and 24 °C, the C–D frequency sharply increased due to the transition to the liquid crystalline state and the concomitant increase in the proportion of *gauche* conformers. Incorporation of a potential drug, SK&F 96079, resulted in a broadening of the transition and a *ca.* 5 °C reduction in its onset. This indicated that the bilayer structure was essentially intact, and the main effect was

[4] R. C. Mitchell and D. C. Lee, in 'Proceedings of the 2nd International Workshop on Fourier Transform Infrared Spectroscopy', ed. E. F. Vansant, Antwerp, 1992, p. 53.

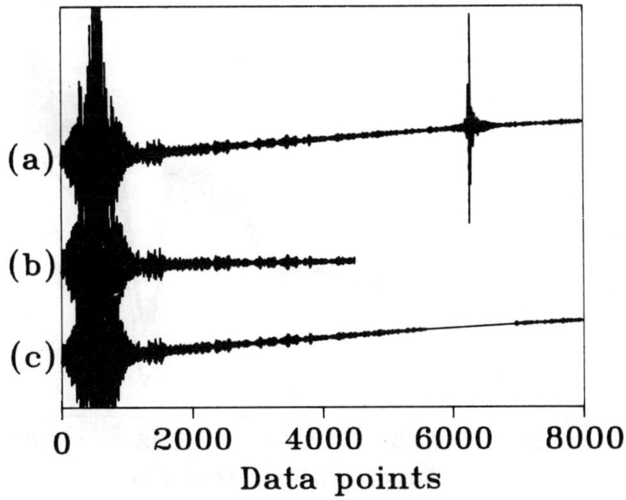

Figure 4 *The interferograms of a silicon wafer.* (a) *Interferogram recorded at* 1 cm^{-1} *resolution. Note the secondary centreburst.* (b) *Same wafer recorded at* 8 cm^{-1} *resolution.* (c) *The same interferogram as in* (a) *but with the secondary centreburst replaced by zeros*

a reduction in the cooperativity of the gel to liquid crystalline phase transition.

The availability of the data in digital form greatly simplifies this type of analysis. Most operating systems allow the programming of small routines, to semiautomate the production of plots such as Figure 3.

2.2.2 Silicon Wafers. IR spectroscopy finds widespread use in the semiconductor industry. The technique is routinely used for quality control purposes to quantify carbon and oxygen in single crystal silicon wafers, to determine the amount of hydrogen in amorphous silicon hydride and to measure the thickness of epitaxial layers.

However, the high refractive indices of the semiconductors causes them to act as beamsplitters. As Figure 4(*a*) shows, this leads to a secondary centreburst in the interogram and when this is processed to give the spectrum, Figure 5(*a*), the result is that the spectrum is dominated by fringes, to such an extent that it is virtually impossible to derive any useful information from the spectrum.

There have been a number of ways that have been used to eliminate the fringes. These include subtracting a sine wave of appropriate amplitude and phase from the spectrum, recording the spectrum at the Brewster angle (where the reflectivity is zero, hence no interference can occur) or recording the spectrum at worse (numerically larger) resolution. The first method is best in restricted ranges of the spectrum because the amplitude varies across the spectrum and this is difficult to model successfully. The second method requires the use of polarized light and fairly accurate positioning of the sample with respect to the incident beam. The last method works because reducing the resolution means that the length of the interferogram is reduced, so the secondary centreburst does not occur in the shorter interferogram, Figure 4(*b*).

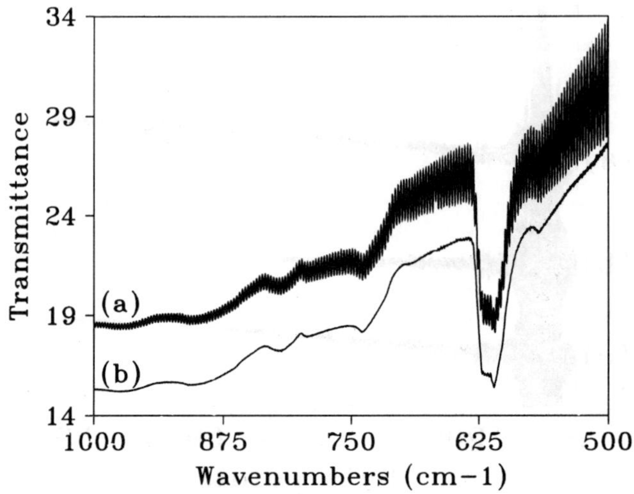

Figure 5 (a) *and* (b) *Spectra generated by Fourier transformation of the interferograms shown in Figure* 4(a) *and* 4(c) *respectively. Note the greatly reduced amplitude of the fringes in* 5(b)

The most successful approach has been to make use of the fact that the data is in digital form and hence readily amenable to manipulation. The process is illustrated in Figure 5. The interferogram is obtained, either directly or by Fourier transformation of the spectrum, and the secondary centreburst located. The portion of the interferogram in the region of the secondary centreburst is then removed and set to zero, Figure 5(*a*), with the result that the fringes are either eliminated or greatly reduced, Figure 5(*b*). More sophisticated versions of this process add noise to the zeroed region or smooth the adjacent regions. In both cases, the aim is to prevent a discontinuity in the interferogram that can introduce ripples into the spectrum. In practice, the simplest method works very well.

The secondary centreburst itself contains valuable information. This is because its position in the interferogram depends on the wafer's thickness, (the interferogram equivalent of using the fringes to measure the pathlength of a cell). If the measurement is carried out in reflection from a sample that consists of an epitaxial layer on a substrate, then the thickness of the epilayer can be derived from the position of the secondary centreburst, providing a non-destructive method for thickness determination.

2.3 Resolution Enhancement

To some degree or other, any technique that outputs the data in a two-dimensional form, be it a spectrum, a chromatogram or a thermogram, suffers from the problem of overlapping peaks. The best way to overcome this problem is to adjust the experimental conditions to eliminate the overlap. If this is not feasible for practical reasons such as the time available or fundamental reasons such as the bands are separated by less than their halfwidths, then the next best solution is to attempt some form of resolution enhancement if the overlap is unacceptable.

There are a large number of methods that have been suggested or employed, but the major ones are: Fourier self-deconvolution (FSD), the maximum entropy method (MaxEnt or MEM), curve-fitting and the use of derivatives. Each of these methods has particular strengths and weaknesses and the first three of these will be illustrated in the next three sections. The reason for the omission of the use of derivatives is that, in the author's opinion, the results are usually so abysmal as to be useless, particularly when compared with the other methods, that the technique has been superseded. The great advantage of the other methods over curve-fitting is that no assumptions about the number of peaks present are required, so this helps eliminate prejudice on the part of the user.

2.3.1 The Oxidation of Polyethylene. IR spectroscopy has been used to characterize the carbonyl compounds formed in thermally and photochemically oxidized polyethylene for over 30 years. The broad carbonyl band typically shows some structure and the major components have been assigned as saturated aldehydes, ketones and acids at 1733, 1721 and 1712 cm^{-1} respectively. However, the bands are severely overlapped and this makes the assignment problematical.

FSD potentially offers a way round this problem. It was used to investigate the products formed by the thermal oxidation of a range of polyethylenes.[5] In FSD there are three choices that need to be made: (*i*) the width of the band that is to be deconvoluted, (*ii*) the degree of resolution enhancement, K, defined as the ratio of the width of the original band divided by that of the deconvoluted band and (*iii*) the apodization function that is used to suppress the oscillations resulting from the use of a finite FT. In this case, preliminary work had shown that the use of a triangular-sinc apodization function gave the best results in terms of signal-to-noise, so this was used throughout. (The implementation of FSD that was used here, was somewhat different from that described in Chapter 1; full details appear elsewhere.[6,7])

The choice of bandwidth and K are interlinked, so the best approach is to vary the bandwidth on a trial and error basis until the bands appear over-deconvoluted, as shown by the presence of negative peaks and then to reduce the width slightly. Once this has been done, a further series of deconvoluted spectra as a function of K can be obtained. The process is illustrated in Figures 6 and 7. In Figure 6(*a*), the spectrum prior to deconvolution shows some indications of components either side of the main peak at 1715.5 cm^{-1}, but there is a dramatic improvement on deconvolution with a bandwidth of 8 cm^{-1} and $K=2.7$, Figure 6(*b*). Four peaks are clearly visible at 1724, 1719, 1715 and 1697 cm^{-1}. As the width is increased, a negative peak at 1726 cm^{-1}, Figure 6(*c*), is indicative of over-deconvolution, although the overall resolution is not impaired. With a bandwidth of 15 cm^{-1}, Figure 6(*d*), the spectrum is clearly over-deconvoluted. This suggests that the component peaks vary somewhat in their halfwidth and demonstrates one of the compromises necessary when FSD is applied to IR spectra: the deconvoluting exponential is only optimized for one of the bands, while for any of the others it is

[5] W.F. Maddams and S.F. Parker, *J. Polym. Sci. B: Polym. Phys.*, 1989, **27**, 1691.
[6] D.I. James, W.F. Maddams and P.B. Tooke, *Appl. Spec.*, 1987, **41**, 1362.
[7] P.B. Tooke, *Trends in Anal. Chem.*, 1988, **7**, 130.

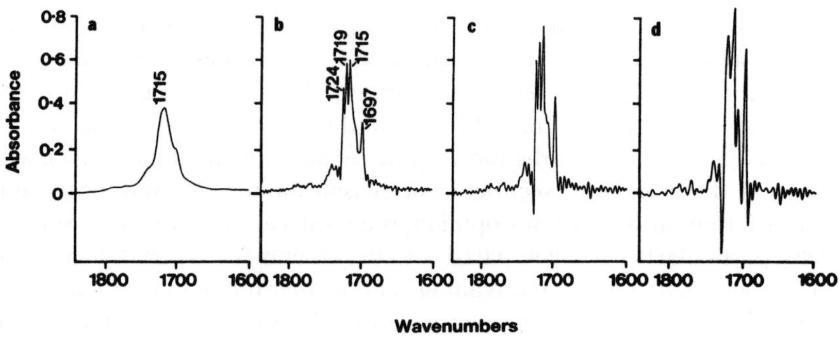

Figure 6 *Spectrum of oxidized high density polyethylene as a function of bandwidth for a* K *value of* 2.7: (a) *original spectrum,* (b) 8 cm^{-1}, (c) 11 cm^{-1} *and* (d) 15 cm^{-1}
From 'Vibrational Spectroscopy of the Oxidation of Polyethylene. I. Fourier Self-deconvolution of the Carbonyl Absorption', W. F. Maddams and S. F. Parker, *J. Polym. Sci.: Part B: Polym. Phys.*, 1989, **27**, 1691, reproduced with permission. Copyright, 1989, John Wiley and Sons, Inc.

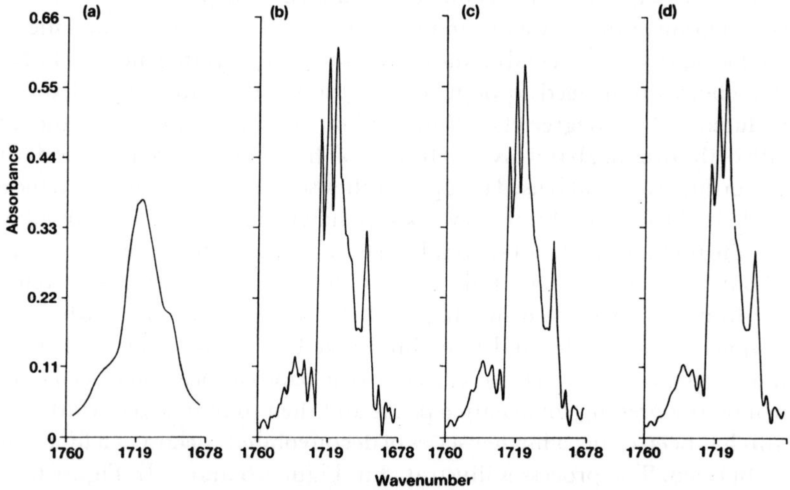

Figure 7 *Spectrum of oxidized high density polyethylene as a function of* K *value of a bandwidth of* 8 cm^{-1}: (a) *original spectrum,* (b) K=3.0, (c) K= 2.7 *and* (d) K=2.5
From 'Vibrational Spectroscopy of the Oxidation of Polyethylene. I. Fourier Self-deconvolution of the Carbonyl Absorption', W. F. Maddams and S. F. Parker, *J. Polym. Sci.: Part B: Polym. Phys.*, 1989, **27**, 1691, reproduced with permission. Copyright, 1989, John Wiley and Sons, Inc.

approximation. From Figure 6, a bandwidth of 8 cm^{-1} was taken as the optimum. Figure 7 shows the result of varying the K value at this bandwidth. It is apparent that the peak positions are virtually independent of K, although as expected, the relative intensities vary somewhat. A comparison of all three K values suggests that the peaks at 1738, 1733 and 1729 cm^{-1}, although weak, are genuine. The background noise level for $K=3.0$ is high and a K value of 2.7 is probably the best

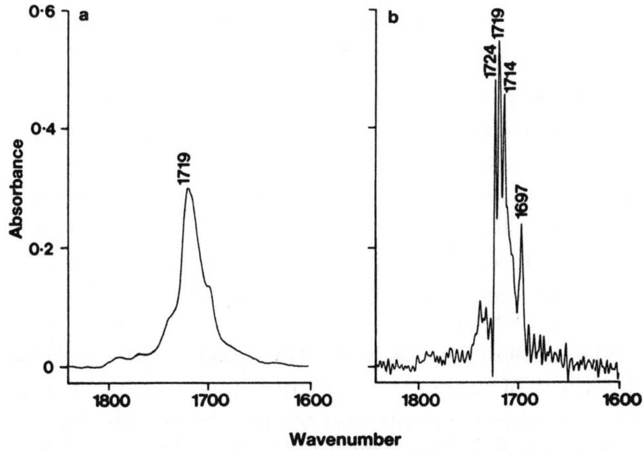

Figure 8 *Spectrum of oxidized high density polyethylene after soaking in concentrated aqueous ammonia*: (a) *original spectrum*, (b) *after deconvolution with a bandwidth of* 8 cm^{-1} *and* K=2.7
From 'Vibrational Spectroscopy of the Oxidation of Polyethylene. I. Fourier Self-deconvolution of the Carbonyl Absorption', W. F. Maddams and S. F. Parker, *J. Polym. Sci.: Part B: Polym. Phys.*, 1989, **27**, 1691, reproduced with permission. Copyright, 1989, John Wiley and Sons, Inc.

choice. FSD is very demanding in terms of signal-to-noise (SNR) in the original spectrum. It may be shown[8] that the maximum K value that can be used is given by:

$$K = \log_{10}(S/N)$$

so in practice it is difficult to justify using a K factor greater than 3.0.

Having located the peaks, the next and more difficult step was to assign them. This was done by a mixture of model compounds and chemistry. Figure 8 shows the same polymer after soaking in concentrated aqueous ammonia. The spectrum, Figure 8(*a*), shows a shift in the peak maximum of Figure 6(*a*), but the deconvoluted spectrum, Figure 8(*b*), shows clearly that the band at 1714 cm^{-1} has decreased in intensity. Together with the appearance of a band at 1500 cm^{-1} in the difference spectrum, this allows the assignment of the 1714 cm^{-1} band to carboxylic acid and the 1550 cm^{-1} band to the carboxylate produced by reaction with ammonia. Comparison with an ethylene–carbon monoxide copolymer, which contains aliphatic ketone groups in the backbone, allows the 1719 cm^{-1} band to be assigned to ketones. Similarly comparison with an ethylene–acrylic acid copolymer[9] suggests that the 1697 cm^{-1} band should be assigned to hydrogen bonded carboxylic acid dimers. Interestingly, this band had not been detected by other studies of oxidized polyethylene and its intensity appeared to be very dependent on the type of polyethylene. The assignment of the band at 1724 cm^{-1}

[8] J. K. Kauppinen, D. J. Moffatt, D. G. Cameron and H. H. Mantsch, *Appl. Opt.*, 1981, **20**, 1866.
[9] M. A. Harthcock, in 'Proceedings of the 1985 International Conference on Fourier and Computerised Infrared Spectroscopy', ed. J. G. Grasselli and D. G. Cameron, Proc. SPIE 553, 1985, p. 245.

remains uncertain; an obvious assignment would be to an aldehyde but these are likely to be transient species so the persistence of the band would be very surprising.

2.3.2 Graphite Intercalates. Graphite is well known for its layered structure of carbon sheets and high electrical conductivity. Intercalation of molecules between the sheets is used to modify the electrical and chemical properties. Amongst the latter are possible uses as catalysts. The goal of this study was to determine the stage of intercalation (the ratio of free graphite layers to intercalated layers).

The Raman spectrum of a graphite/$FeCl_3$ intercalate[10] in the region of the in-plane vibration of the graphite sheets is shown in Figure 9(*a*). The spectrum is characterized by a poor SNR and it is clearly inappropriate to use FSD to enhance the spectum. The result of a maximum entropy reconstruction using a Lorentzian function with a width of 25 cm^{-1} is shown in Figure 9(*b*). The result is dramatic: a noise-free spectrum with two completely resolved peaks. These may be assigned as the in-plane vibration from free graphite layers at 1582 cm^{-1} and the in-plane vibration from the bounding intercalated graphite layers at 1617 cm^{-1} and allows the material to be assigned as a stage 3 intercalate.

The intercalated materials are highly reactive and deterioriate rapidly in oxygen, even at the parts per million level that leak into 'airtight' sample holders. Figure 10(a) shows the spectrum recorded five hours after Figure 9 and 10(*b*) the result of the MEM reconstruction. The 1581 cm^{-1} band has decreased in intensity and the feature due to intercalation has broadened and shifted to 1613 cm^{-1}. This indicates that the sample has undergone oxidation to a stage 3/4 or stage 4 intercalate.

The great strength of MEM is that no assumptions about the number of components are required and it is capable of producing spectacular results from very noisy data. The disadvantages are that it is conceptually more difficult to understand than either FSD or derivatives and that the results are very dependent on knowing the correct form and width of the blurring function. Thus MEM works best where all the peaks are the same width and shape, such as in nuclear magnetic resonance (NMR) spectroscopy.

MEM has also suffered from a lack of general availability, in contrast to FSD which is available on virtually all FTIR spectrometers as part of the standard software package. This situation has changed now that a low cost implementation of MEM and the conceptually similar maximum likelihood method is available for a PC.[11]

2.3.3 Crystallinity of Polyethylene. The simplest model of polyethylene is of a three-phase system: a crystalline phase, a liquid-like amorphous phase and a semi-disordered interfacial phase. The major structural factors that affect the performance of polyethylene with regard to its mechanical properties, such as strength, toughness and resistance to cracking, are the molecular weight and its distribution, the level and type of branching, the density and the crystallinity.

[10] S. Davies, K. J. Packer, A. Baruya and A. I. Grant, in 'Maximum Entropy in Action, ed. B. Buck and V. A. MaCauley, Clarendon Press, Oxford, 1991.

[11] SSRes from Spectrum Square Associates. The program is described in *American Laboratory*, March 1990 and August 1991.

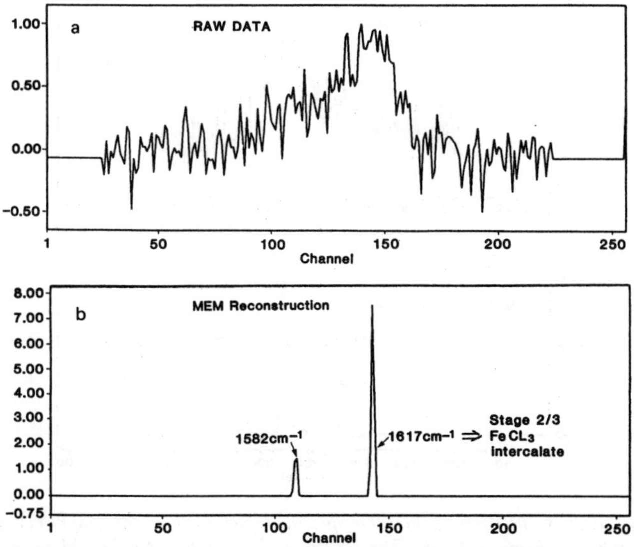

Figure 9 (a) *Raman spectrum of a graphite*/$FeCl_3$ *intercalate in the in-plane stretching region of graphite around* 1580 cm^{-1}. (b) *MEM reconstruction with a Lorentzian function of* 25 cm^{-1} *width*
S. Davies, K.J. Packer, A. Baruya and A. I. Grant, in 'Enhanced Information Recovery in Spectroscopy Using the Maximum Entropy Method' from 'Maximum Entropy in Action', eds. B. Buck and V. A. MaCauley. Clarendon Press, Oxford, 1991, p. 73, by permission of Oxford University Press.)

The crystallinity may be measured by a variety of techniques including DSC, X-ray diffraction and from the density. In 1978 Strobl and Hagedorn[12] proposed a method for measuring the crystallinity from the Raman spectrum. This has a number of advantages, principally because the measurement is localized rather than an average from the entire sample. With a Raman microscope samples may be as small as 1 μm diameter, meaning that polyethylene fractions or inclusions could be measured.

The method has recently been re-evaluated with state-of-the-art instrumentation, using both dispersive and FT Raman spectroscopy, and modern computing techniques.[13] Figure 11 shows FT Raman spectra from three samples of different densities and crystallinities. The essence of the procedure is that the crystalline fraction is determined from the CH_2 bending mode at 1415 cm^{-1}, the amorphous fraction from either the broad CH_2 twisting mode at 1303 cm^{-1} or the C–C stretching band at 1080 cm^{-1} and the interfacial component is determined by difference. The individual contributions are measured by curve-fitting the relevant regions of the spectrum and using the integrated intensity of the CH_2 twisting modes at 1296 and 1303 cm^{-1} to normalize to unity. Figure 12 shows an example of a curve-fit spectrum.

[12] G. R. Strobl and W. Hagedorn, *J. Polym. Sci., Polym. Phys. Ed.*, 1978, **16**, 1181.
[13] K. P. J. Williams, S. M. Mason, N. Conroy, C. Naylor, R. Meier and B. J. Kip, manuscript in preparation.

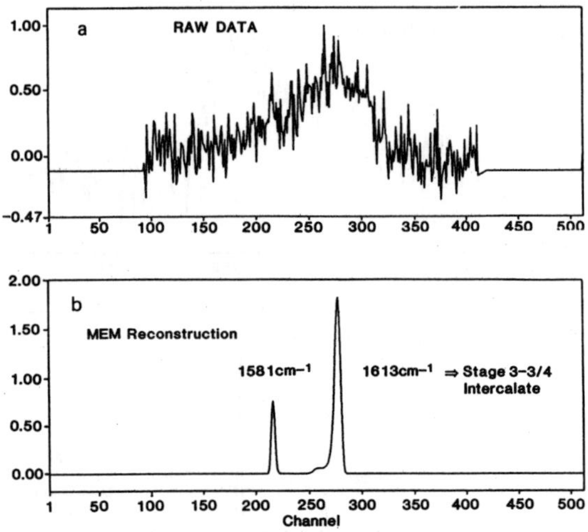

Figure 10 (a) *Raman spectrum of the same spectral region of the same graphite*/$FeCl_3$ *intercalate sample as shown in Figure* 7, *recorded five hours later.* (b) *MEM reconstruction with a Lorentzian function of* 25 cm^{-1} *width*
S. Davies, K. J. Packer, A. Baruya and A. I. Grant, in 'Enhanced Information Recovery in Spectroscopy Using the Maximum Entropy Method' from 'Maximum Entropy in Action', eds. B. Buck and V. A. MaCauley. Clarendon Press, Oxford, 1991, p. 73, by permission of Oxford University Press.)

Strobl and Hagedorn found a good correlation between the crystallinity as measured by X-ray diffraction and that determined from the 1415 cm^{-1} band. The recent work found that the correlation was poor at low crystallinities and proposed a crystallinity measurement based on the 1060 cm^{-1} band. As shown in Figure 13, an excellent correlation was obtained. The reason for the improvement is that the 1415 cm^{-1} band represents the presence of perfectly ordered material, as found for high density polyethylene. As the density decreases, the perfection of the ordered phase is diminished so intensity in the Raman spectrum is lost, but is still maintained in the X-ray measurement which measures order over a larger scale. The 1060 cm^{-1} band is due to short, 4–6, segments of ordered *trans* methylene groups. As such, it is less sensitive to small disruptions of crystallinity and provides a better correlation with the X-ray data at low crystallinities.

2.4 Multi-technique

To some extent, the examples in the previous sections have been misleading, in that they relied on a single technique to provide the answer. In industry, it is much more common that the problems are so complex, that no single method will provide the full story, and hence a variety of spectroscopic, optical and physical tests have to be used.

2.4.1 Customer Complaints. A fairly common source of complaints in the polymer industry is of defects in the raw polymer, as pellets or film, or the finished product.

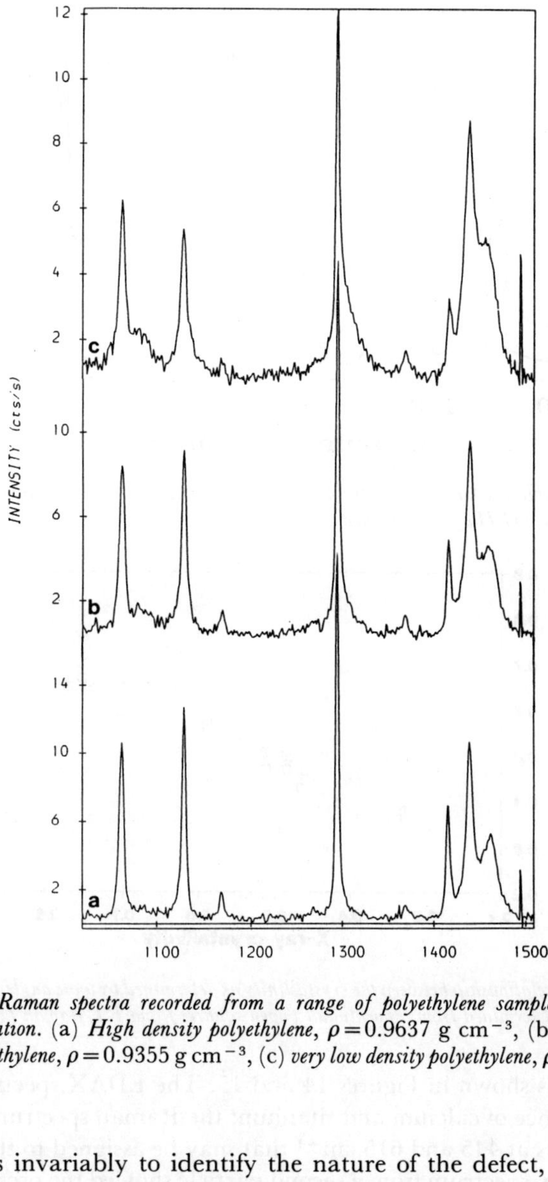

Figure 11 *FT Raman spectra recorded from a range of polyethylene samples using* 1064 nm *excitation.* (a) *High density polyethylene,* $\rho = 0.9637$ g cm^{-3}, (b) *linear low density polyethylene,* $\rho = 0.9355$ g cm^{-3}, (c) *very low density polyethylene,* $\rho = 0.9035$ g cm^{-3}

The first task is invariably to identify the nature of the defect, whether it is a different polymer, an undispersed additive, a catalyst residue or dirt. The contaminants range in size from millimetres to sub-micron.

One approach that has proven particularly useful is to examine the sample in a scanning electron microscope (SEM) and use energy dispersive X-ray analysis (EDAX) to obtain an elemental analysis. This allows the contaminant to be located and because of the micron spatial resolution of EDAX, by checking several areas, it is possible to deduce whether there is more than one contaminant. The sample may then be examined by vibrational spectroscopy to give molecular characterization.

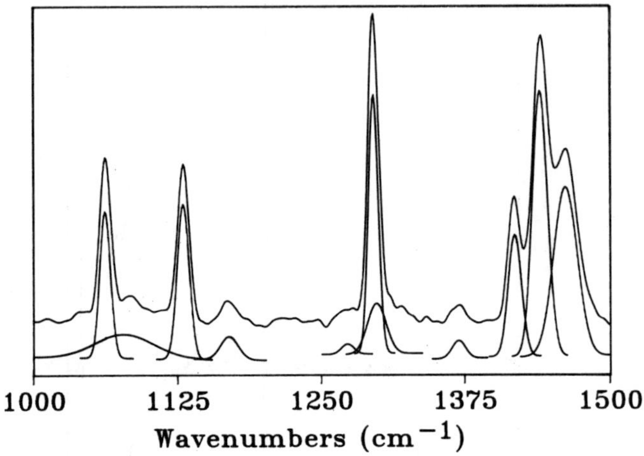

Figure 12 *Example of a curve-fit spectrum for the determination of crystallinity in polyethylene by the Strobl and Hagedorn method*

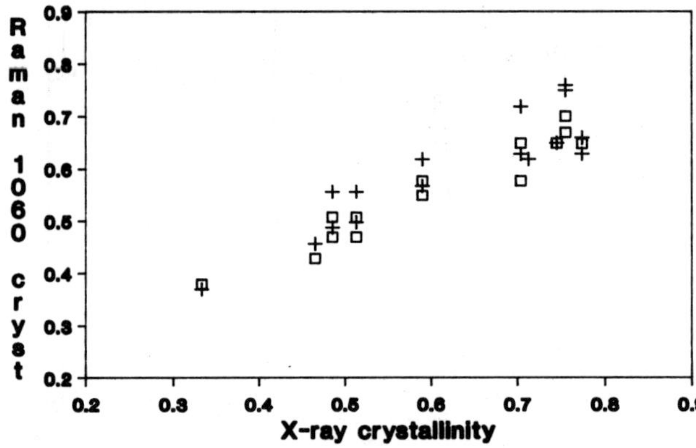

Figure 13 *The relationship between the crystallinity as determined by wide angle X-ray scattering and that determined from conventional (open squares) and FT Raman (crosses) spectroscopy*

An example is shown in Figures 14 and 15. The EDAX spectrum, Figure 14, shows the presence of calcium and titanium; the Raman spectrum, Figure 15(*a*), shows two bands at 445 and 615 cm^{-1} that may be assigned to the rutile form of TiO_2. An EDAX spectrum from a second particle showed the presence of chlorine and the Raman spectrum, Figure 15(*b*), showed the presence of C–Cl bands at 635 and 695 cm^{-1} and two bands at 1125 and 1515 cm^{-1} that can be assigned to the C–C and C=C stretch of a conjugated polyene chain. Together this strongly suggests the presence of degraded polyvinyl chloride (PVC). Calcium carbonate and titania are common fillers in PVC and contamination by filled PVC seems to be the cause of the problem.

2.4.2 Water-trees in Polyethylene. One of the modes of failure of the cross-linked polyethylene used for cable insulation is by water-treeing. Water-trees are

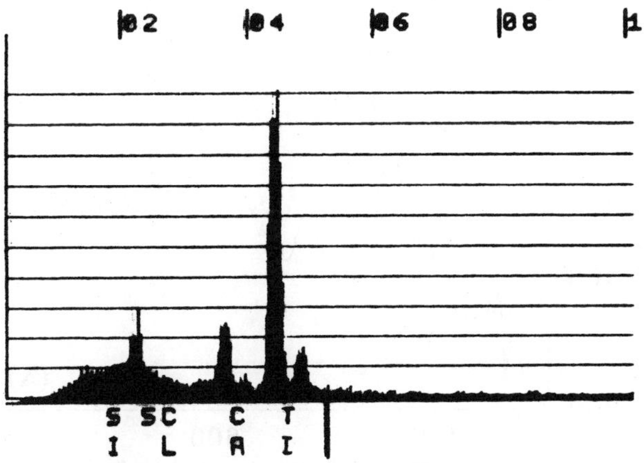

Figure 14 *EDAX spectrum of a polymer contaminant*

damaged areas that when boiled in water and viewed with an optical microscope have a dendritic appearance. The mechanisms of formation are not well understood, and in order to develop effective water-tree retardant polyethylene insulation it is necessary to be able to define the types of chemical species present and their spatial distribution.

Water-trees are typically of the order of a few millimetres or less, and thus microscopical techniques are employed. One of the newer optical techniques is confocal laser scanning microscopy (LSOM). This technique is described in detail elsewhere[14] but in essence it eliminates the problem of limited depth of field. At a given height, the sample is scanned with a laser and only the in-focus light is allowed to reach the detector. By repeating this process at varying heights, it is possible to build up an image in the computer memory where all of the image is in focus. Figure 16 shows an LSOM image of a water-tree.[15] This enables the spatial extent to be defined but does not give any information on the species present.

Elemental information on the micron scale can be obtained by proton induced X-ray emission (PIXE); Figure 17 shows the distribution of potassium in the water-tree (note the close correspondence to Figure 16).

Information on the molecular species present can be obtained by IR microscopy of microtomed sections. Figure 18(*a*) shows the spectrum of undamaged polyethylene and Figure 18(*b*), the water-treed polyethylene. A band at 1600 cm^{-1} is clearly present in the water-treed specimen and can be assigned to carboxylate ions. By the use of a computer controlled XY microscope stage, spectra can be recorded from the area containing the water-tree and a map of the carboxylate distribution derived by ratioing the height of the 1600 cm^{-1} band to a

[14] A basic description of confocal microscopy is given in: A. Hall, M. Browne and V. Howard, *Proc. Royal Microscopical Soc.*, 1991, **63**, and a more mathematical approach in: C. J. R. Sheppard and T. Wilson, *Optik*, 1980, **5**, 331.

[15] W. B. Reid, L. J. Rose, V. Rose, A. Campus, C. Baker and J. R. Walton, in 'Proceedings of the Sixth International Conference on Dielectric Materials, Measurements and Applications', 1992, p. 496.

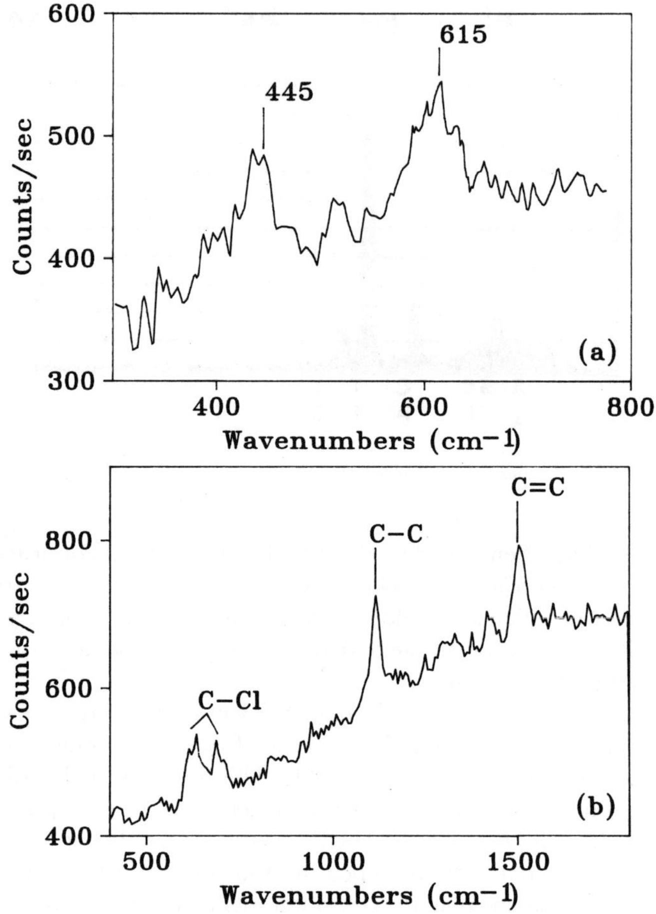

Figure 15 *Raman spectra of* (a) *the contaminant analysed by EDAX in Figure* 14 *and* (b) *a different type of contaminant from the same sample*

polyethylene band at 2020 cm^{-1} to eliminate the effects of any variation in thickness across the section. The result is shown in Figure 19. The concentration of carboxylate was higher to the base of the tree and decreased out from it. This provides a plausible partner for the potassium observed using PIXE and suggests that some stage of the formation of water-trees involves oxidation of the polymer.

3 CONCLUSIONS

In this chapter, the intention has been to demonstrate some of the ways in which computers are routinely used in industrial vibrational spectroscopy. There are two major omissions: the first is the use of chemometrics which is increasingly being used for both laboratory and 'on-site' plant applications; this is dealt with elsewhere in this book. The second is the use of combined techniques particularly gas chromatography infrared spectroscopy, GC-IR, which although not as widely

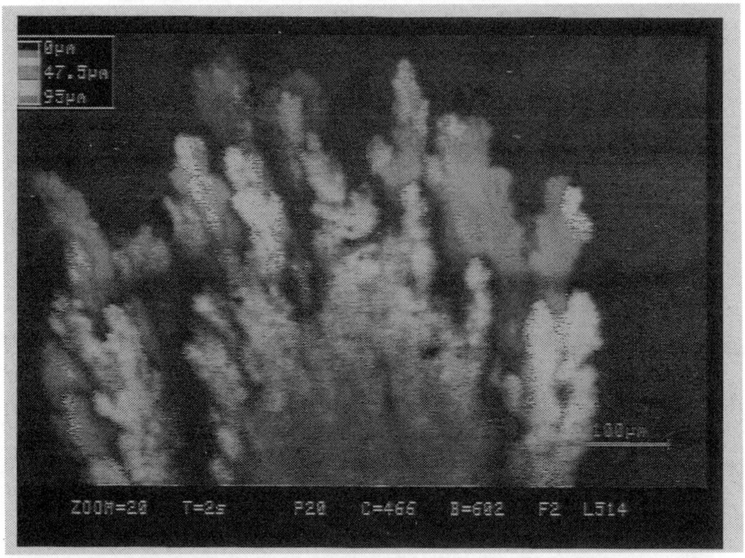

Figure 16 *Fluorescence LSOM image of a water-tree*

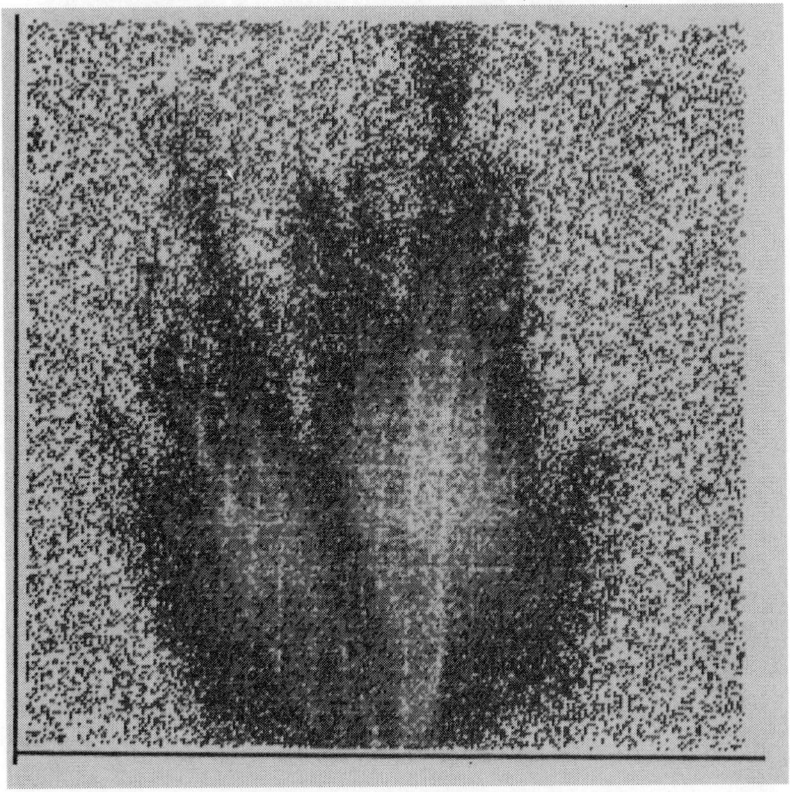

Figure 17 *PIXE image of the distribution of potassium in the water-tree shown in Figure* 16

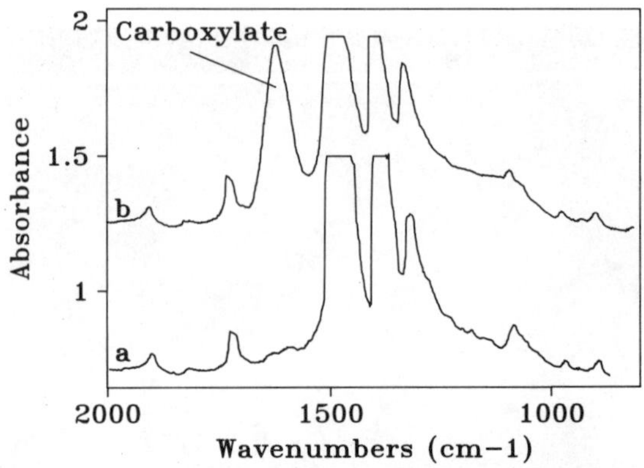

Figure 18 *IR spectra of polyethylene cable insulation.* (a) *Undamaged area and* (b) *water-tree area*

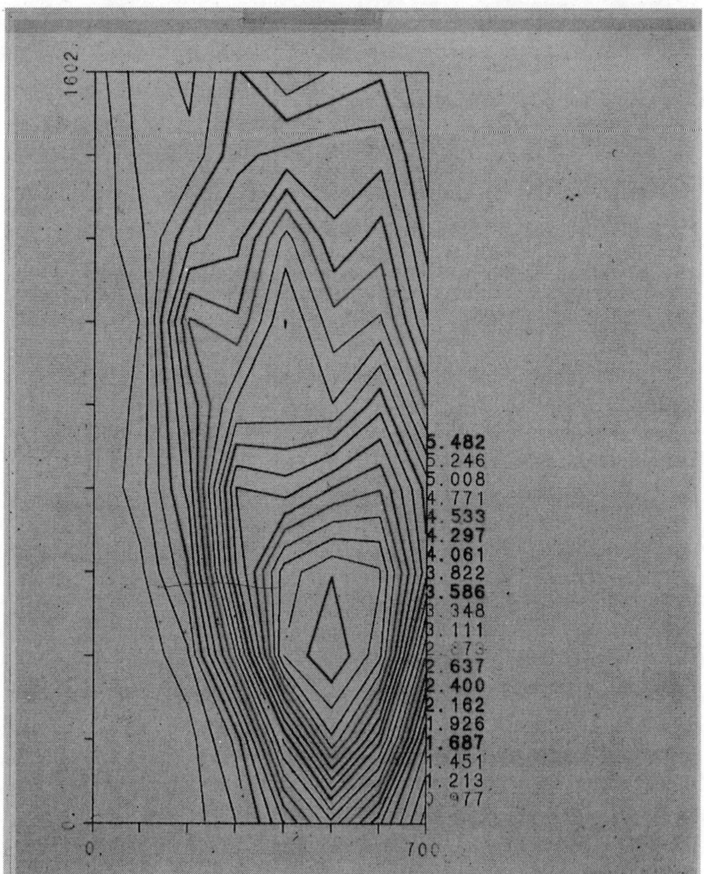

Figure 19 *Contour plot of the distribution of carboxylate ions in the water-tree*

used as GC-MS, has niche applications in industries such as the perfume and pharmaceutical industries: anywhere that identification of isomers is critical. In addition, there are demonstration or commercially available systems for techniques ranging from thin layer chromatography–IR to supercritical fluid chromatography–IR to thermogravimetric analysis–IR. With the emphasis on increased information per analysis these are areas that will move into the mainstream of vibrational spectroscopy.

4 ACKNOWLEDGEMENTS

The author would like to express his thanks and appreciation for helpful discussions to Clive Baker and also for providing some of the figures and to Ken Williams for providing a copy of ref. 13 prior to publication. Permission to publish has been given by BP International plc.

CHAPTER 9

Development and Applications of Raman Microline Focus Spectrometry

D. J. GARDINER

1 BACKGROUND

Over the last 20 years Raman microscopy has become established as a significant analytical tool in both academic research environments and in industry. Early instrument designs set out to obtain Raman data from an area on the sample surface and to produce species specific Raman images. However, instrument and detector limitations rendered the observation of Raman images very difficult in all but a few idealized samples. Single point instruments on the other hand were applied successfully to many areas of research and analysis.[1] Applications include a study of working elastohydrodynamic bearing contacts,[2] the formation of microcrystals in human joint tissue,[3] works of art,[4] fibres,[5] corrosion products[6] and stress measurements.[7]

Development of the single point instrument to allow collection of data from arrays of points on a surface under computer control has opened up the possibility of realistic Raman imaging. An example is provided by the study of stress induced phase changes in stabilized, tetragonal, zirconia ceramic.[8] Figure 1 shows the Raman spectra of both tetragonal and monoclinic zirconia. Raman spectra from an area on the surface of a sintered zirconia ceramic tile were recorded at 3 μm intervals over an area 210 μm × 210 μm. The data were processed to yield the intensity ratio of the monoclinic bands at 181 and 192 cm^{-1} to the tetragonal band at 264 cm^{-1} as a grey scale and to plot this as a function of position. Figure 2 shows the resulting image in which the regions of high monoclinic content appear black.

[1] 'Practical Raman Spectroscopy', ed. D. J. Gardiner and P. R. Graves, Springer, Berlin, 1989.
[2] D. J. Gardiner, E. M. Baird, C. Craggs, M. P. Dare-Edwards and J. C. Bell, *Lub. Sci.*, 1989, **1**, 301.
[3] N. McGill, P. A. Dieppe, M. Bowden, D. J. Gardiner and M. Hall, 'The Lancet, 1991, **337**, 77.
[4] S. B. Best, R. J. Clark and R. Withnall, *Endeavour*, 1992, **16**, 66.
[5] N. J. Everall and J. Lumsdon, *J. Mater. Sci.*, 1991, **26**, 5269.
[6] D. H. Gardiner, C. J. Littleton, K. M. Thomas and K. N. Strafford, *Oxidation of Metals*, 1987, **27**, 59.
[7] J. Birnie, C. Craggs, D. J. Gardiner and P. R. Graves, *Corrosion Sci.*, 1992, **33**, 1.
[8] M. Bowden, G. D. Dickson, D. J. Gardiner and D. J. Wood, *J. Mater. Sci.*, 1993, **28**, 1031.

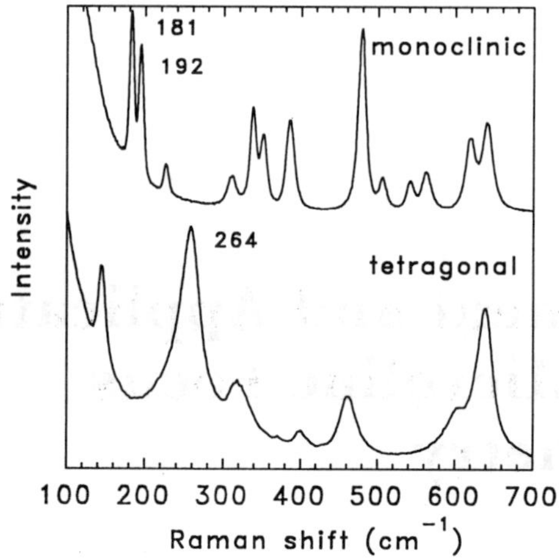

Figure 1 *Raman spectra of tetragonal and monoclinic zirconia* (ZrO_2) (Adapted from ref. 8)

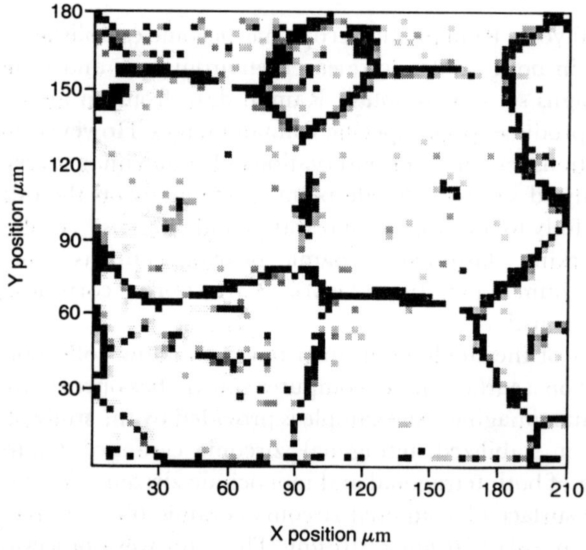

Figure 2 *Raman image of a sintered zirconia ceramic; regions of high monoclinic content appear black* (Adapted from ref. 8)

The pattern of black lines coincides exactly with the grain boundaries in the ceramic tile. A smaller region of the same tile was also mapped after being stressed with a diamond indenter. The resulting image is shown in Figure 3 where significant transformation to the monoclinic form has occurred in the region of the indent. The small rectangular grain at the top of the region studied serves to register the two images. A further example is provided by a study of inherent stress

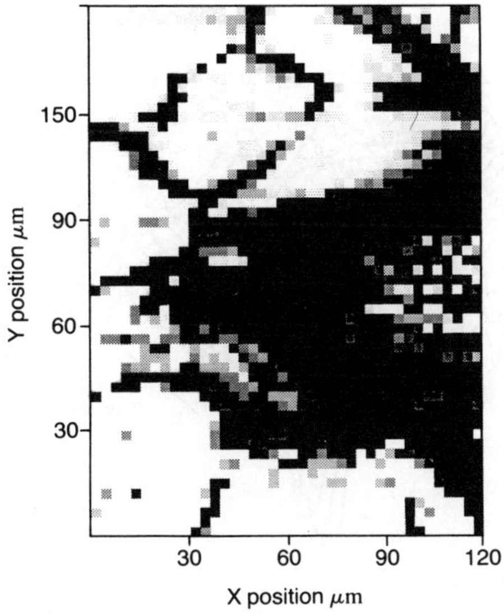

Figure 3 *Raman image of part of the sintered zirconia ceramic shown in Figure 2, following stressing by diamond indentation. Regions of high monoclinic content appear black* (Adapted from ref. 8)

in silicon devices. Here the frequency of the optical photon of silicon at 520 cm^{-1}, which is sensitive to stress, was measured as a function of position[9] over an area on an advanced bipolar VLSI transistor employing trench isolation technology. The data shown in Figure 4 was collected at 0.5 micron intervals and shows the resulting stress surface which reveals significant stress concentration around the trenches. A commercial system, developed by Dilor, allows mapping to be carried out without moving the sample. This is achieved by using two movable, coupled, lenses to control the position of the laser spot on the sample and simultaneously to make this point conjugate with the monochromator slit. This system, though more complex optically, has an obvious advantage over the simpler approach in the study of samples, which are not easily moved. In addition, only course mechanical movements of the coupled lenses are required for a given displacement of the point of analysis, a significant feature when displacements of the order of 1 μm are required.

Point-by-point mapping has several advantages over imaging by global illumination, not least because the total laser power available is utilized to record each set of data and complete spectra are recorded for each point which can be analysed in a variety of ways after the experiment. In addition the method can be conducted under strictly confocal conditions allowing the possibility of three-dimensional imaging to be tackled. However, the advent of the slow scanned charge coupled device (ccd) detector, with its improved sensitivity and

[9] W. J. Rothwell, C. Hill and P. J. Pearson, 'Proc. XIIIth. Int. Conf. Raman Spectrosc.', Wurzburg, Germany, 1992, p. 1064.

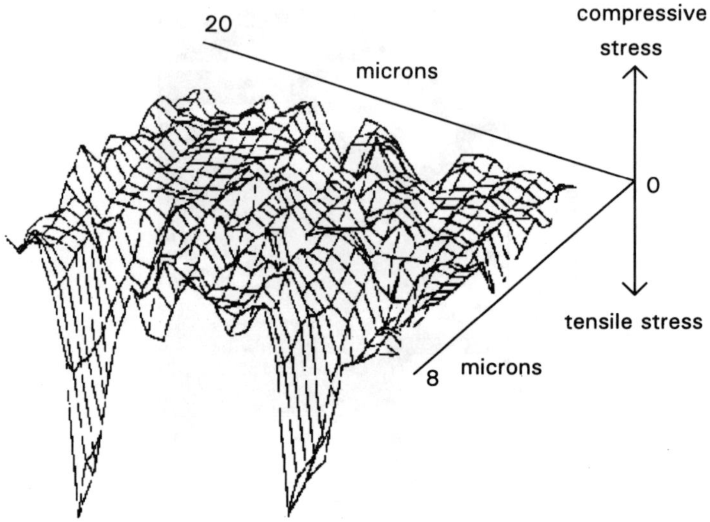

Figure 4 *Stress surface generated from Raman shift changes in the optical phonon of silicon at* 520 cm^{-1} *measured from point-by-point mapping of an advanced bipolar VLSI transistor* (Adapted from ref. 9)

two-dimensional character, opens up new imaging and profiling possibilities.[10–12]

The ccd is a silicon device comprising potential wells in a two-dimensional array with no intervening dead space. Incident photons with energy higher than the silicon band gap produce electron hole pairs with high quantum efficiency (35%). The photoelectrons are stored at their incident locations in the potential well array whilst the holes are swept from the device. The resulting charge image is read out by the application of suitable clock waveforms to electrodes on the ccd chip. This moves one row of charge at a time (in the x direction) into the output register where it is then moved in the y direction, one unit at a time to an on-chip amplifier. The output voltage is directly proportional to the number of electrons in the potential well unit. In this way the whole array of potential wells (pixels) can be read out *via* an analogue to digital converter (ADC) and presented as a video image on a computer monitor. When operated at liquid nitrogen temperatures, the device has very low (essentially zero) dark current. In addition the read out noise is small, resulting in a detector which is capable of producing good signal/noise spectra from the very low light level sources encountered in Raman spectroscopy. The use of back-thinned chips and low noise multi-pin phase (MPP) chips, which can be used effectively with Peltier cooling, are now being routinely used for Raman studies.

[10] S. B. Dierker, C. A. Murray, J. D. Legrange and N. E. Schlotter, *Chem. Phys. Lett.*, 1987, **137**, 453.
[11] D. N. Batchelder, *Eur. Spectrosc. News*, 1988, **80**, 28.
[12] M. Bowden, D. J. Gardiner, G. Rice and D. L. Gerrard, *J. Raman Spectrosc.*, 1990, **21**, 37.

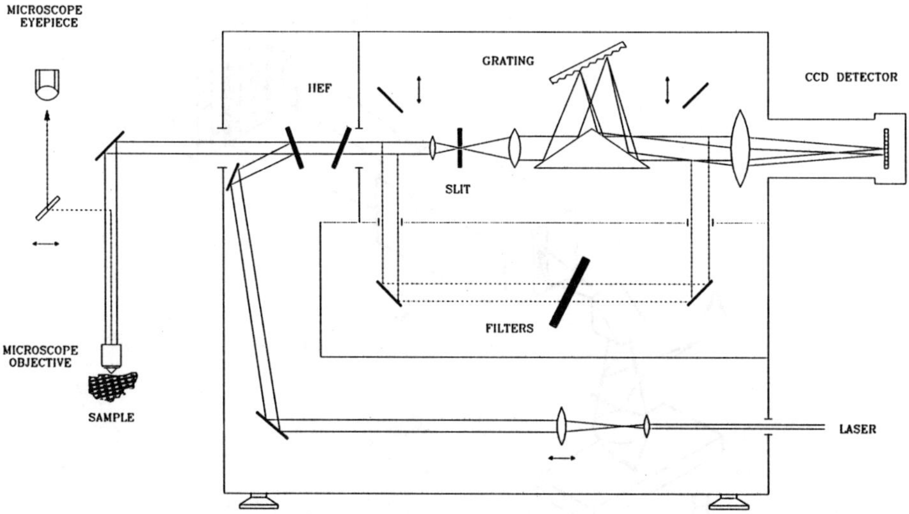

Figure 5 *Schematic of global illumination Raman microscope* (Ramascope figure supplied by D. N. Batchelder)

1.1 Raman Profiles and Images

There are two basic approaches to producing Raman profiles, maps or images with ccd detectors.

1.1.1 Global Illumination. Here the whole of the area of interest on the sample is illuminated by the laser beam either by rastering the beam or by defocusing the laser spot on the sample. The scattering light is then imaged through a variable filter, which selects the required spectral region, onto the ccd detector. The filter may be either a non-astigmatic spectrograph or a tuneable narrow band pass filter. This approach has been pioneered by Batchelder *et al.* and the resulting instrument is primarily an imaging device but also having limited spectroscopic potential. It is able to work both in the single point or global illumination (imaging) mode.[13] A schematic of the device is shown in Figure 5. Holographic edge filters (HEF) are used to reflect the laser beam into the objective and the scattered light is then filtered of laser light by the same, plus an additional, HEF. The Raman scattered and any fluorescence light is then either dispersed by a simple grating spectrograph to generate a Raman spectrum, or is filtered to select a chosen Raman band for imaging. The use of HEFs improves greatly the overall transmission of the instrument but limits the spectroscopy close to the exciting line. Each exciting line will also require its own set of filters.

1.1.2 Line Scan or Focus. The approach developed by ourselves is to produce a line focus on the sample and to image the scattered light from the line focus through a non-astigmatic filter/spectrography system onto the ccd detector. In this way Raman spectra from all points along the line focus are analysed simultaneously and a Raman profile is generated. If required, Raman map data can be obtained

[13] A Garton, D. N. Batchelder and C. Cheng, *Appl. Spectrosc.*, in the press.

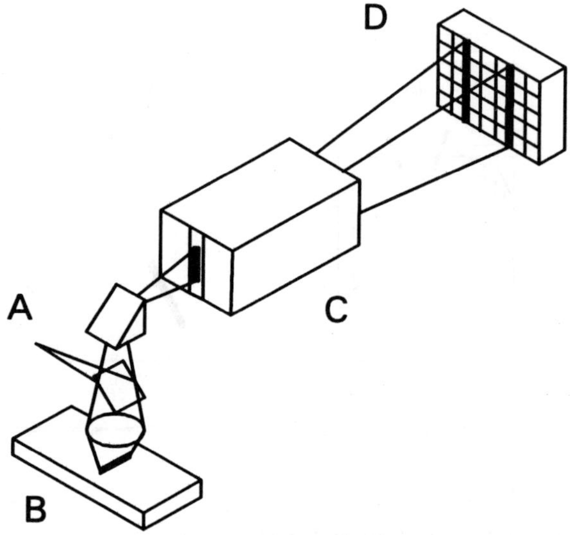

Figure 6 *Principle of Microline Focus Spectrometry (MiFS).* (A) *laser in,* (B) *sample,* (C) *filter/spectrograph,* (D) *ccd*

by moving the sample orthogonally to the analysis line. Figure 6 shows a schematic of the system for which we have coined the acronym MiFS – microline focus spectrometry. An alternative method which is being pursued by the Dilor company involves a modification of their spot scan technique to provide a linear scan of the laser spot on the sample surface. Once again the integrity of the Raman scatter from each spot position is retained through the filter and dispersion system and is captured on a ccd detector.

2 RAMAN MICROLINE FOCUS SPECTROMETRY (MiFS)

The principal components of the Raman MiFS instrument are shown in Figure 7. The laser beam is introduced into the microscope objective *via* a beam splitter (B1) and a cylindrical lens (C). The latter, produces a line focus on the sample, of length 50–1000 μm, dependant upon the magnification of the objective. Lenses (L1–L6) are used to collimate and focus throughout the system in order that the optical path can be on axis, thus eliminating astigmatism, which would otherwise distort the resultant images and profiles. The first two gratings (G1, G2) work in subtractive mode to filter out the laser wavelength, with an intervening inverter (not shown). The Raman spectrum is then dispersed by grating G3 onto the ccd detector. The sample and focused laser are viewed simultaneously using a video camera (V) by means of the movable beamsplitter (B2). White light is provided from the lamp house (W). The cylindrical lens is rectractable allowing rapid interchange from line to spot focus. It is further possible to bypass the filter stage and to employ edge filters when working away from the exciting line or when transmission is critical. In addition to Raman profiles, automatic imaging is

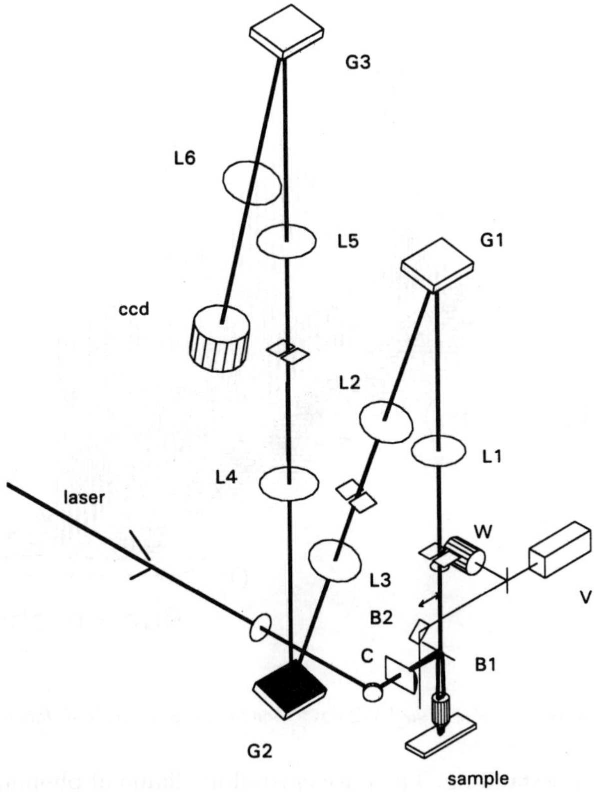

Figure 7 *Schematic of Raman MiFS; see text for detail*

achieved by use of a programmed sample stage capable of 0.06 μm steps. The MiFS control software allows data collection and processing in real time to produce Raman images from any selected combination of spectroscopic parameters. In this way images showing for example, species concentrations, orientation, and phase change can readily be generated.

3 RAMAN MiFS APPLICATIONS

The microline focus geometry has distinct advantages over the single point configuration in many respects. These include, fast profiling and mapping, coherent spatially resolved data, representative average data, low power density at the sample and dynamic range maximization. The following applications of the technique highlight some of these aspects.

3.1 Diamond Films

There has been an explosion of Raman interest in carbon and diamond since it has been possible to realize the production of synthetic diamond films both on

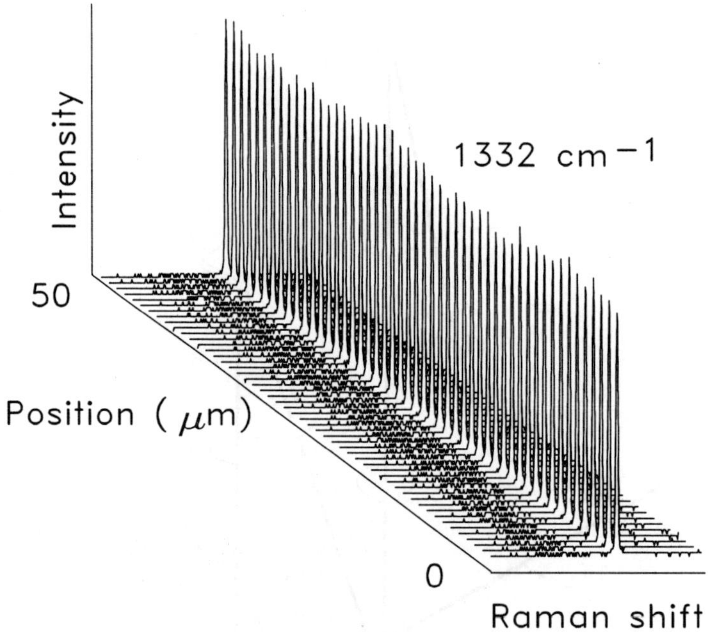

Figure 8 *Raman MiFS profile of the* 1332 cm^{-1} *band obtained from the surface of a gem diamond*

substrates and free standing. The pure crystalline diamond phonon which gives rise to an intense Raman band at 1332 cm^{-1} can be used to indicate quantity, quality and inherent stress of the diamond. Other forms of carbon including graphite, disordered graphite and amorphous material also are strong Raman scatterers thus establishing Raman spectroscopy as one of the principal tools for the study of diamond and carbon films. Figure 8 is a MiFS profile of the 1332 cm^{-1} band obtained from the surface of a gem diamond. The small changes in intensity along the line arise from contamination on the diamond surface. A very different MiFS profile is shown in Figure 9, which was recorded from an acetylene flame deposited diamond film. The data show variations in the Raman spectra along a 50 μm analysis line with 2 μm spatial resolution. There is clear evidence of variations in graphite carbon characterized by bands between 1400 and 1600 cm^{-1} and the presence of species giving rise to a band at around 1200 cm^{-1}. Summing (binning) all of the Raman data along the analysis line results in the average Raman spectrum shown in Figure 10. Analysis of this type of data is valuable in providing not only details of the spatial distribution of carbon species within the film, but also for rapidly providing a representative average set of data to give an overall impression of film quality.

3.2 Amorphous Carbon Films

Amorphous carbon films can be prepared by rf-plasma, ion-beam and microwave deposition methods. Amorphous carbon is known to anneal to the disordered

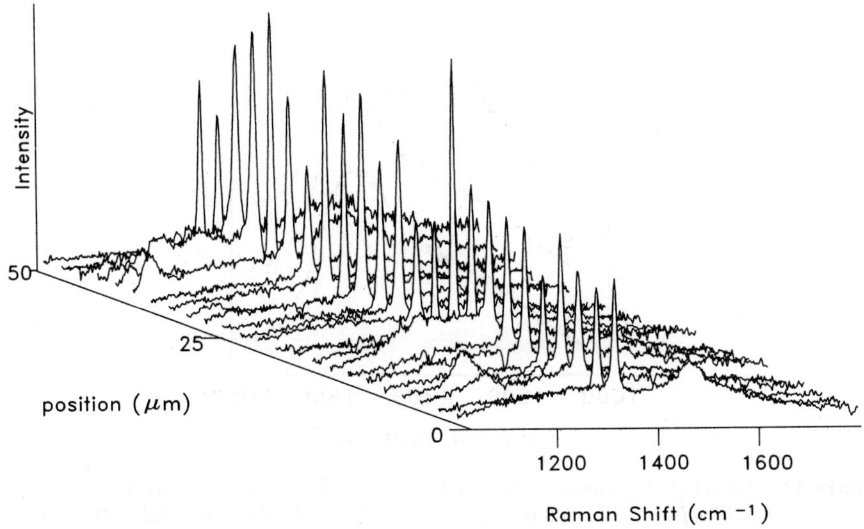

Figure 9 *Raman MiFS profile of an acetylene flame deposited diamond film*

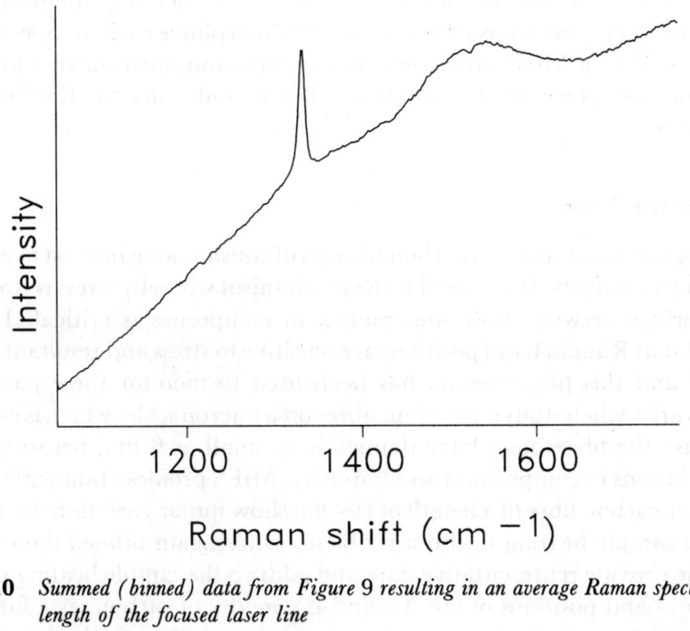

Figure 10 *Summed (binned) data from Figure 9 resulting in an average Raman spectrum from the length of the focused laser line*

graphite phase at 600 °C. This makes Raman studies difficult with the high power densities of single point instruments. Figure 11 shows the Raman spectra recorded[14] from an ion-beam deposited amorphous carbon film using a point focus

[14] M. Bowden, D. J. Gardiner and J. M. Southall, *J. Appl. Phys.*, 1992, **71**, 521.

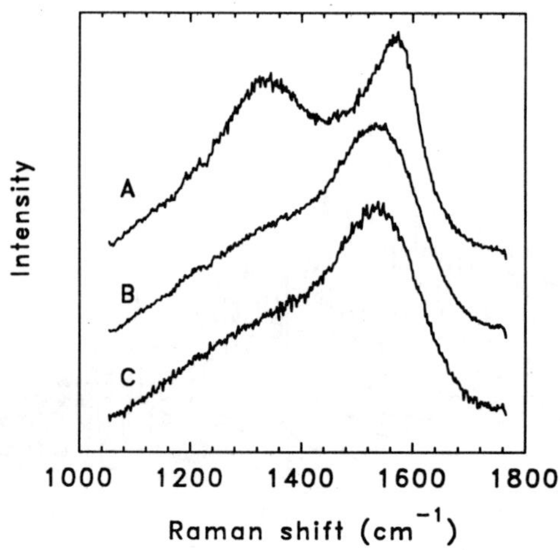

Figure 11 *Raman spectra from an ion-beam deposited a C–H film using a* 30 mW *line focus with an* X20 *objective* (C), *an* 8 mW *spot focus with an* X20 *objective* (A) *and a* 0.8 mW *spot focus with an* X20 *objective* (B)
(Adapted from ref. 14)

(A), which shows the two distinct features of the disordered graphite spectrum, and a line focus (C), which is characteristic of amorphous carbon. Spectrum (B) was obtained from a MiFS instrument working in point focus mode but only by reducing the laser power at the sample to 0.8 mW and collecting the data over a long exposure.

3.3 Carbon Fibres

Monitoring stress and strain in carbon fibres is of considerable interest to engineers and materials scientists. In particular the mechanism whereby stress is transferred at the interface between fibre and matrix in composites is critical. It is well established that Raman band positions are sensitive to stress and resultant strain in a material and this phenomenon has been used to monitor these parameters. Difficulties arise when studying carbon fibres on a micron scale, which is essential if only because the fibres may have diameters as small as 6 μm, because of local surface variations in composition and impurity. MiFS profiles obtained[15] from a 6 μm diameter carbon fibre of a length of 100 μm show minor variations in intensity, in addition sample heating is also a problem. Once again binned data over the analysis line provide representative data and address the sample heating problem. The Raman band positions of the A_{1g} and E_{2g} modes of carbon, as a function of laser power at the sample, for an unsized, pitch-based fibre, for both point focus and MiFS line focus geometries, are plotted in Figure 12. The MiFS data is unaffected by laser power variations over the range used whereas the point focus data incurs shifts of the same magnitude as those induced by stress. Using the MiFS

[15] M. Bowden, D. J. Gardiner, J. M. Southall and D. L. Gerrard, *Carbon*, in the press.

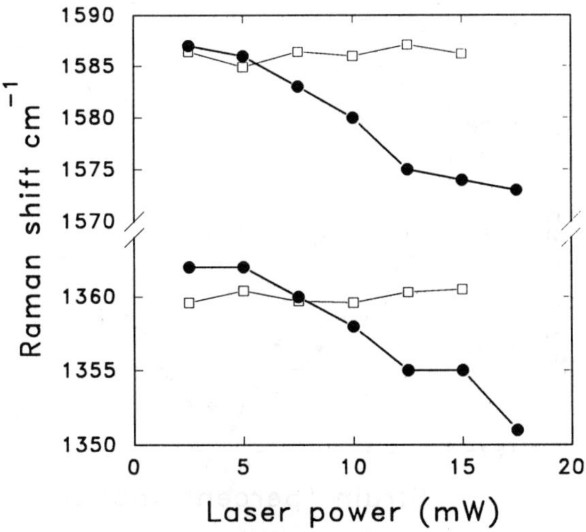

Figure 12 *The effect of laser power at the sample, on the Raman band positions of the A_{1g} and E_{2g} modes of a carbon fibre; □ line focus data; ● point forcus data*
(Reproduced from ref. 15 with permission)

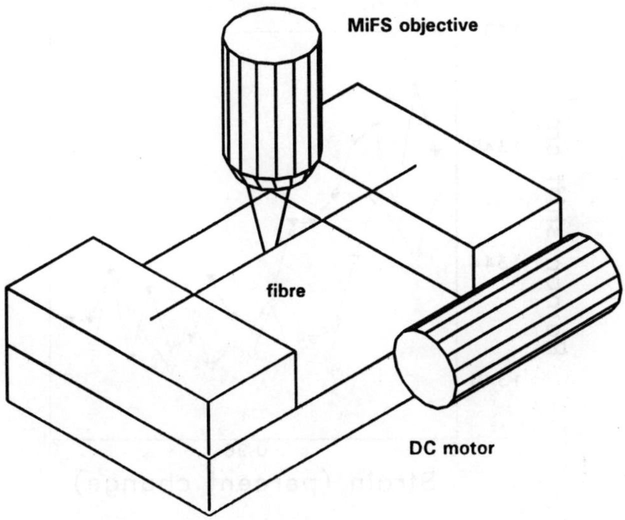

Figure 13 *Carbon fibre stress rig and MiFS geometry*

approach, Raman spectra were obtained from carbon fibres under the application and release of stress using a simple rig shown schematically in Figure 13. The resulting data for two fibres from the same tow are displayed in Figures 14 and 15 and demonstrate the differing response of the A_{1g} and E_{2g} modes for these fibres under these conditions. The reduced frequency of the A_{1g} band after the straining cycle is probably due to an irreversible change in the microstructural disorder of the graphite plane edges.

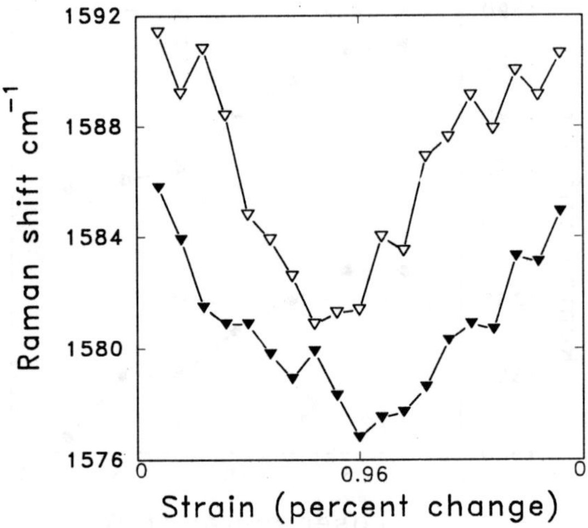

Figure 14 *The effect of increasing and decreasing strain on the E_{2g} Raman band position for two carbon fibres from the same tow*
(Reproduced from ref. 15 with permission)

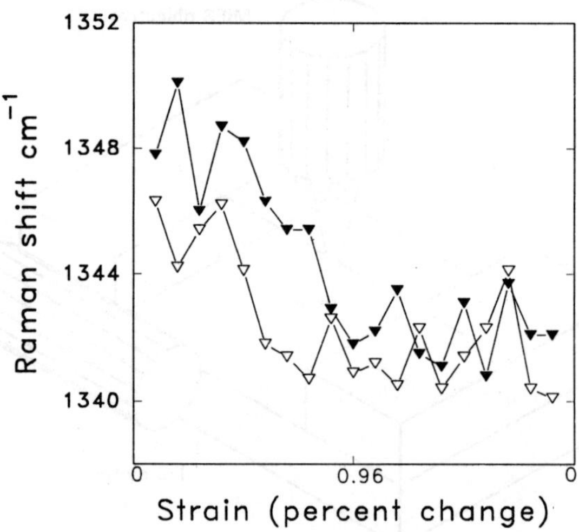

Figure 15 *The effect of increasing and decreasing strain on the A_{1g} Raman band position for two carbon fibres from the same tow*
(Reproduced from ref. 15 with permission)

3.4 PVC Degradation

Polyvinylchloride is known to degrade at elevated temperatures and by exposure to UV radiation and chemical attack, through loss of hydrogen chloride. This loss results in the formation of chains of doubly-bonded carbon atoms in the polymer, which can be detected at very low concentration by resonance Raman spectroscopy.

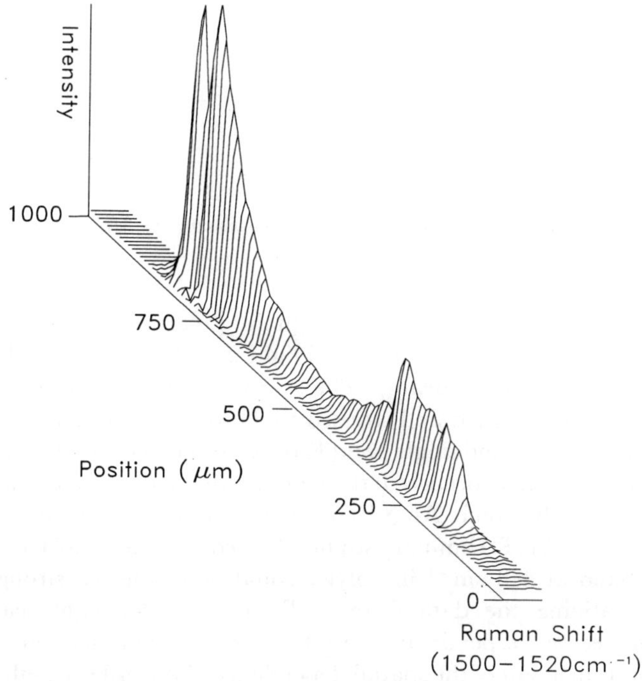

Figure 16 *Raman MiFS profile of the resonance Raman bands arising from the polyene degradation products across the sheet thickness of polyurethane backed polyvinylchloride*

It is of interest to determine the rate at which the degradation proceeds, the extent to which the effect will ingress into the polymer sheet and the effect of the presence of polyurethane backing. Experiments were devised[16,17] in which polyurethane backed polyvinylchloride sheet was thermally degraded. The sheet was then sectioned and analysed by Raman MiFS. Figure 16 is a MiFS profile of the resonance Raman bands arising from the polyene degradation products across the sheet thickness. The result provides clear evidence of the extent of degradation ingress and of the additional degradation occurring at the polyurethane foam side as a result of attack from residual amines in the foam.

3.5 Stress Mapping in Polycarbonate

Stress patterns in polycarbonate have been investigated using Raman MiFS.[18] Polarized Raman intensity is dependent upon molecular orientation as determined by the Raman scattering tensor. A Polaroid sheet, placed in the optical path from

[16] M. Bowden, P. Donaldson, D.J. Gardiner, J. Birnie and D. L. Gerrard, *Anal. Chem.*, 1991, **63**, 2915.
[17] J.W. Bradley, D.J. Gardiner, J. Birnie, N.M. Dixon, D.L. Gerrard and A.S. Wilson, 'Proc. XIIIth. Int. Conf. Raman Spectrosc.', Wurzburg, Germany, 1992, p. 1048.
[18] M. Bowden, J.W. Bradley, D.J. Gardiner and B.J. Kip, 'Proc. XIIIth. Int. Conf. Raman Spectrosc.', Wurzburg, Germany, 1992, p. 1046.

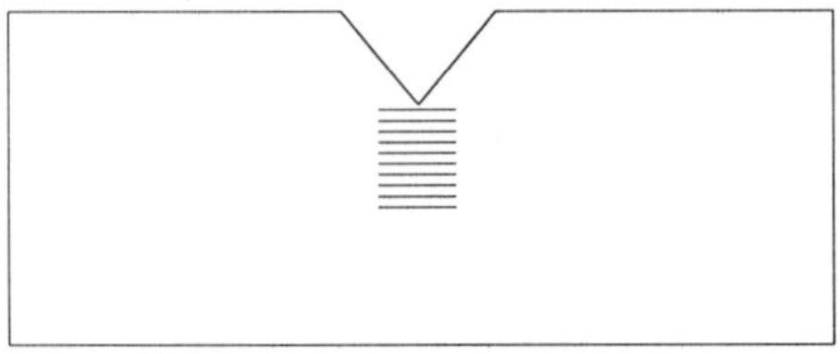

Figure 17 *Schematic of notched polycarbonate sample, showing the MiFS mapped region*

sample to spectrometer was used to obtain polarized Raman spectra (*i.e.* measurement of the intensity of scattered light parallel and perpendicular to the polarization of the exciting beam). The sample used was a 3 mm thick IZOD polycarbonate strip notched and stressed to 33 MPa resulting in yielding at the notch tip. Birefringence and single point Raman studies indicated that appreciable molecular ordering had occurred throughout the sample, particularly in the region at the tip of the notch. An X10 objective was used, with a total laser power at the sample of 5 mW. Preliminary studies showed that the depolarization ratio of the intense band at 888 cm^{-1} in polycarbonate was affected strongly by stress variations. Ratioing the data from MiFS profiles, for light scattered with polarization vector perpendicular (to the incident polarization vector) and parallel, results in a plot of the spatial dependence of the ordering effect along the analysis line. Such a plot shows depolarization ratios of 0.1–0.2 in regions of random molecular orientation; however, in the regions where molecular orientation had occurred the ratio varies from 0.7 to >1.0 with the degree of orientation. Polarized Raman data were collected from a series of focused lines on the surface of the sample as depicted in the schematic of Figure 17. The mapped area was 500 μm (the length of the microline) by 200 μm, at intervals of 25 μm. The 888 cm^{-1} bands were integrated to give a single intensity value from each of the 250 points along the line then, for simplicity, every ten data points were summed to give 25 data points along the line. The result, shown in Figure 18, is a map of the variation of the depolarization ratio and hence molecular orientation across the surface of the sample. Within the limited area studied there is a clear indication of regions of orientated polymer radiating out from the notch tip.

3.6 Dynamic Range

An interesting consequence of line focusing coupled with ccd detection, is that a quasi extension of the dynamic range of the detector can readily be achieved compared to the spot focus geometry. A line focus spreads the laser power along the line such that more ccd pixels are used to detect the light. As a result the total number of photons which can be detected before saturation occurs is increased. For homogeneous samples a line focus will produce the same spectral data as a spot focus, but binning up the resultant line data in software following readout, allows the sum of the full dynamic range of each of the pixels detecting scatter from the analysis line to be utilized. This has important implications in dealing with highly

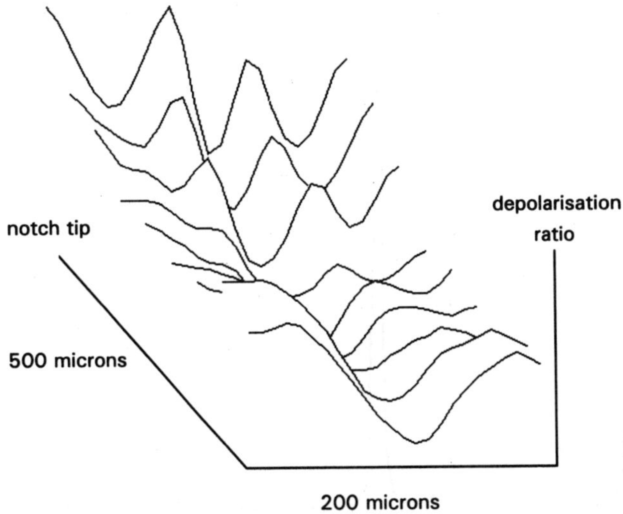

Figure 18 *Raman map of the variation of the depolarization ratio, and hence molecular orientation across the surface, of a stressed polycarbonate sample, in the region shown in Figure* 17

fluorescent samples which frequently saturate a detector before acceptable signal/noise data can be obtained.

3.7 Species Imaging

An example of Raman imaging is provided by a MiFS instrument test, designed to demonstrate the production of images from Raman scatter close to the exciting line. Figure 19 is the MiFS Raman spectrum of Hg_2Cl_2 showing the 41 cm^{-1} Raman shift band. Using the instrument in its mapping mode, with the computer controlled stage moving in 0.25 μm steps, the near real time image (a few seconds following each line exposure are required for data processing) shown in Figure 20 was produced from some grains of Hg_2Cl_2 using the 41 cm^{-1} Raman scatter with background subtraction.

4 THE FUTURE

There are two distinct avenues for future instrument development. In the first place, improved general research instrumentation would combine the ability of multi-element detectors to acquire high quality micrographic data sets in reasonable time, with convenient access to the full range of spectroscopic measurements needed to determine all the available physical parameters exemplified in this article. In contrast, the identification of specific measurement applications having widespread commercial or research relevance could lead to dedicated instruments designed for mapping a single pre-determined spectroscopic

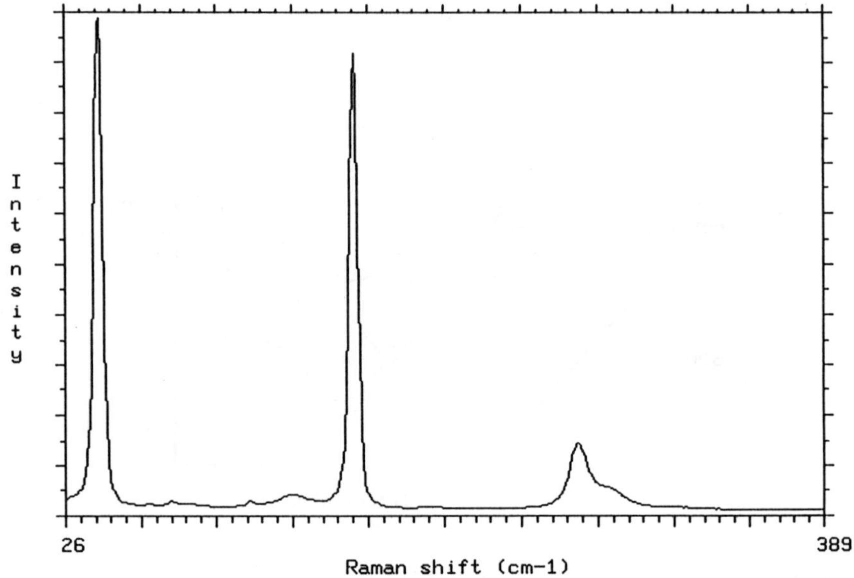

Figure 19 *Raman MiFS spectrum of* Hg_2Cl_2 *grains using point focus*

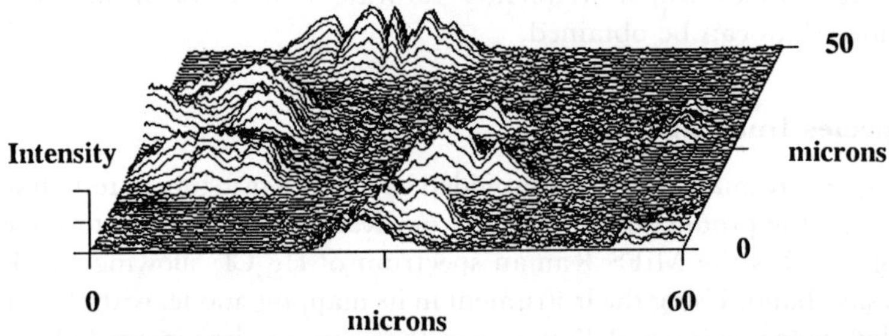

Figure 20 *Raman MiFS map of* Hg_2Cl_2 *grains using the* 41 cm^{-1} *Raman shift band*

parameter. Both the MiFS and global illumination approaches have contributions to make in these respects. In addition the possible applications of a Raman imaging system based upon an acousto-optic tuneable filter are being energetically pursued.[19]

[19] E. N. Lewis, P. J. Treado and I. W. Levin, 'Proc. XIIIth. Int. Conf. Raman Spectrosc.', Wurzburg, Germany, 1992, p. 1072.

CHAPTER 10

Near Infrared Spectroscopic Techniques

G. DOWNEY

1 INTRODUCTION

Near infrared (NIR) spectroscopy has become widely adopted for rapid analysis in the food, agricultural, pharmaceutical and chemical industries over the last 10–15 years. There are very many reasons which explain this popularity but perhaps the most significant include its ease of operation, the requirement for only minimal (or even no) sample preparation, and the ability to analyse material which contains significant amounts of water. Continuous improvements in instrument design and performance have underpinned the reliability of the technique, especially in manufacturing environments, and online industrial applications are becoming increasingly common. Parallel developments in mathematical data treatments have improved the accuracy and stability of calibrations for quantitative work while developing the exciting field of qualitative analysis.

This short chapter will expand on these and related issues to provide an overview of this spectroscopic technique; for further information, the reader is referred to more detailed texts.[1–3]

2 PRINCIPLES AND PRACTICALITIES

2.1 Theoretical Background

The NIR spectral region is normally understood as encompassing the 1100–2500 nm (9091–4000 cm^{-1}) range; of recent times, attention has also focused on the 750–1100 nm (13333–9091 cm^{-1}) region which has been variously described as the near near infra-red (NNIR) or the Herschel IR.[4]

[1] 'Near Infrared Spectroscopy in Food Analysis', ed. B. G. Osborne and T. Fearn, Longman Scientific and Technical, Harlow UK, 1986.

[2] 'Near Infrared Technology in the Agricultural and Food Industries', ed. P. Williams and K. Norris, AACC Inc., St. Paul, USA, 1987.

[3] 'Handbook of Near-Infrared Analysis', ed. D. A. Burns and E. W. Ciurczak, Marcel Dekker Inc., New York, 1992.

[4] A. M. C. Davies, *Appl. Spectrosc.*, 1990, **44**, 14A.

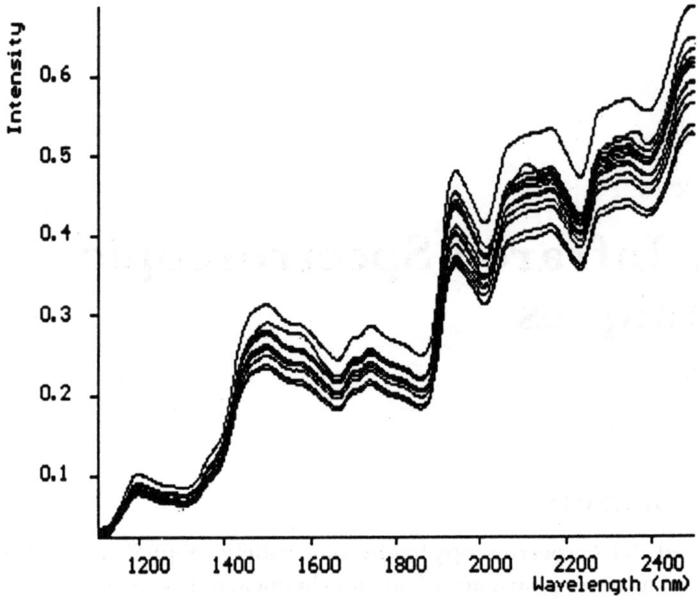

Figure 1 *Collection of spectra of commercial skim-milk powder samples*

Absorption bands seen in the NIR are overtones and/or combination bands of fundamental vibrations occurring in the mid-IR spectral region; these arise from vibrational and rotational movements of covalent bonds. Overtones (which arise because the fundamental vibrations are not purely harmonic in nature) occur at approximately integer multiples of the frequency of the fundamental. Overtone intensity decreases with increasing order and those higher than the third order are of little practical significance.

Overtones commonly encountered in the NIR region arise from fundamental absorption bands occurring at 3 to 8 microns (3000–10 000 nm; 3333–1250 cm^{-1}). These are owing to vibrations between light atoms joined by strong covalent forces – in practice this usually means between hydrogen (H) and either carbon (C), oxygen (O) or nitrogen (N). Bonds of this type are common in organic material and the resulting multiplicity of overtones gives rise to the typical, relatively ill-defined NIR spectrum shown in Figure 1. A further consequence of this number of bond types is that no single wavelength may be identified to which only one component in a mixture responds. In practice this means that measurements have to be taken at a number of wavelengths to perform quantitative analyses.

Combination bands may occur as a result of the interaction of two or more vibrations to produce bands which are the sum of multiples of their fundamental frequencies. They may only arise between different vibrational modes of the same functional group; these modes must also possess the same symmetry.

Accurate calculation of the exact location of overtone or combination bands in the NIR region is compromised by approximations in the theoretical model used (*e.g.* uncertainty in the value of the anharmonicity constant) and the variable effect of the molecular environment of absorbing species (*e.g.* polar, steric and H-bonding effects). Despite this difficulty, a number of spectral tables attributing absorptions at various wavelengths to more or less specific molecular species have been published;[1,2] some degree of caution must be used in the interpretation of real spectra, however, especially in relation to combination band assignments. A recent development, which should underpin the validity of these ascriptions, is in the correlation of NIR and mid-IR spectra of the same compounds;[5] while this work is in its early stages and has focused on complex biological systems rather than simpler molecular species, it holds significant promise.

2.2 Operational Advantages and Disadvantages

The intensity of overtone bands is normally several orders of magnitude lower than the corresponding fundamental absorptions. Each extra quantum of vibration usually involves a reduction in intensity of a factor of about 10. NIR measurements are therefore made on weakly absorbing materials. Operationally this is an advantage over spectroscopy in the mid-IR as it permits the use of convenient sample thicknesses; for example, in the 800–1100 nm wavelength region, NIR transmittance may be measured through 10 cm of water or oil, 2 cm of potato or whole grain, 1 cm of homogenized meat or cream and 0.3 cm of powders such as flour.[6] Additionally, it means that useful spectra can be taken in the presence of water. A disadvantage, however, is that the actual differences in, for example, reflectance values arising from small changes in analyte concentration are very small indeed; a 0.1% difference in protein content between two samples of wheat produces a change in the optical parameter measured (reflectance) of 0.002 units at the most sensitive wavelength. This level of change has required the development of instrumentation with a high signal-to-noise ratio (SNR) and the use of internal standardization procedures in order to achieve acceptable measurement precision. The proximity of the emission maximum of the light source used in NIR instruments to the spectral region of interest, with the consequent advantages of high energy throughput and low stray light, together with the use of solid state detectors has resulted in typical SNRs of 10 000 : 1 in dedicated NIR equipment.

Sample cell construction is normally of relatively inexpensive optical materials such as glass or quartz which are transparent to NIR radiation. A recent development which contravenes this generalization is the use of fibre optics for remote measurement; depending on the spectral region under examination, these fibres can cost upwards of several thousand $US per metre although as the number and range of applications increases, it is expected that this cost will reduce significantly.

Probably the major disadvantage of NIR spectroscopy lies in its empirical

[5] F. E. Barton, II and D. S. Himmelsbach, in 'Near Infra-red Spectroscopy. Bridging the Gap between Data Analysis and Applications', ed. K. I. Hildrum, T. Isaksson, T. Naes and A. Tandberg, Ellis Horwood, 1992, pp. 35–40.

[6] K. H. Norris, 'Multivariate analysis of raw materials' in 'Chemistry and World Food Supplies: the New Frontiers'. Chemrawn II, ed. L. W. Shemilt, Pergamon Press, 1983, pp. 527–535.

nature. Most NIR applications are performed on granular material by a reflectance technique; the effects of light scatter, which are considerable and as yet incompletely modelled, complicate spectral collections by producing vertical offsets between different samples (and even between the same sample of different particle sizes – see Figure 1). Additionally, the spectral baselines are distorted since the magnitude of scattering interferences increases with increasing wavelength. Such is the importance of scattering effects that most of the variation found in any NIR spectral collection may be attributed to this phenomenon rather than to compositional differences.

In the absence of a rigorous mathematical description of the interaction of light with scattering media, a calibration must be developed between the molecular species of interest and the optical signals recorded from a set of real samples. Calibration development and maintenance is a tedious procedure and, unless undertaken correctly, will fail to produce robust equations; failure to predict accurately unknown samples has in the past been responsible for considerable disenchantment with this technique but procedures to maximise the stability of derived calibrations are now quite well-defined. (This topic is more fully dealt with below.)

Given the overlapping nature of NIR spectra referred to above, multivariate statistical techniques are required for calibration development. Practical application of NIR spectroscopy therefore had to await the development of microcomputers and suitable software packages.

3 INSTRUMENTATION

The light source commonly used in NIR instruments is an inexpensive quartz halogen tungsten bulb operating under a constant voltage. Used in conjunction with a spherical mirror behind the bulb, a collimating lens system and a chopper wheel, this produces a pulsed source of light. Silicon detectors are utilized for the 400–1100 nm spectral range while lead sulfide is the material of choice for 1100–2500 nm.

Detector location has been a problem in NIR instrument design and a number of approaches have been used; common to all has been the desire to interfere as little as possible with the amount of light incident on the sample while collecting the maximum quantity of emerging radiation. For transmittance measurements, the detector location is obvious but must be as close as possible to the sample. For reflectance readings, the main consideration is to collect the maximum amount of diffusely-reflected radiation and the minimum of specular reflectance. A number of strategies have evolved:

(*a*) multiple stationary detectors above the sample at an angle of 45° to the incident light with or without a revolving sample cup arrangement;
(*b*) location of two detectors at 45° to the incident light mounted on the interior wall of an integrating sphere coated with a diffuse-reflectance standard such as gold. This limits sample presentation to the reflectance mode but facilitates good measurement stability;
(*c*) as (*a*) above but with a stationary sample cup and precessing detectors.

Wavelength selection techniques vary according to the level of instrumentation involved. In simpler equipment, this has been facilitated by the use of either fixed interference filters or rotating filter paddles although these latter are no longer of any commercial significance. Fixed interference filters permit a high energy throughput with a band pass of around 10 nm at 50% peak transmittance, which is adequate for most NIR applications. A number of these (typically 4–20) are mounted in a metal wheel that rotates to insert the relevant filters sequentially and normally to the light path. This approach is relatively simple and robust, the major long-term problem relating to dust build-up on the filters. Additionally, the possibility exists to add extra filters after the initial purchase of the instrument as the range of applications (and therefore necessary wavelengths) increases.

In the case of tilting filters, the interference filters are attached to a rotating paddle; as the angle of the incident light on the filter changes, so also does the wavelength of the refracted light. Thus, as the paddle turns, each mounted filter scans a small portion of the NIR region. Using a system of encoders, the instrument software is able to track the wavelength of the emerging light from a knowledge of the precise position of the paddle at any time. Such instruments normally held 3–7 filters. One manufacturer utilized this process exclusively but no such instruments are now produced commercially.

Monochromator instruments have traditionally been used mainly for research purposes although this is now changing as costs fall and the analytical requirements of certain industries become more complex. Traditionally, such instruments were based on a moving holographic grating and scanned the range of 1100–2500 nm, but in the recent past this wavelength range has extended downwards to additionally cover the range 400–1100 nm. At least one dedicated NIR manufacturer has produced a Fourier transform (FT-NIR) instrument, which operates using sliding crystal wedge technology. This eliminates the need for moving mirrors and is claimed to be vibration-free, thus allowing for use of the equipment in industrial environments. The extreme wavelength accuracy of FT systems may be of some advantage in the analysis of pure chemical compounds. Another development has been aimed at generating wavelengths without the need for a grating, the movement of which is claimed to mitigate against high wavelength accuracy. A commercial solution to this problem has been the use of an acousto-optic tunable filter (AOTF). AOTF technology uses a monocrystal of tellurium dioxide (TeO_2) on a thin, piezo-electric monocrystal plate. When a high frequency electric signal is applied to this plate, it generates an acoustic wave which propagates through the TeO_2; if a polarized light beam impinges simultaneously on the crystal face at right angles to the direction of the acoustic wave, interference occurs between the two. Exit optics allow the selection of discrete wavelengths and variation of the acoustic wave frequency allows the generation of wavelengths across the NIR spectral range. Again, the main benefit claimed for this system is the complete absence of moving parts with consequent ruggedness and wavelength accuracy.

Novel instruments based on photodiode array detectors,[7] charge-coupled

[7] E. Stark, K. Luchter, Macro- and Micro-specimen Analysis by Near-infrared Diode Array Spectroscopy. Presented at the 4th International Conference on Near Infrared Spectroscopy, Aberdeen, 1991.

devices[8] and Hadamard transform spectrometers[9] have been reported and prototype instruments have been developed. Some interesting applications of this and other technologies in miniature (hand-held) instruments have been described[10] but only limited information exists on their performance.

4 CALIBRATION DEVELOPMENT AND EVALUATION

Calibration development is made on a discrete set of samples and can be a significant undertaking both in terms of time and effort. Essentially it consists of a number of more-or-less well-defined steps which, if adhered to, maximize the likelihood of derived calibrations being robust *i.e.* producing accurate results for unknown samples over a significant time period. Calibration development can be tedious, especially in the case of natural biological materials but any attempt to compromise on this step is almost certain to lead to poor predictive performance.

Given the empirical nature of NIR spectroscopy, the performance of calibrations must be evaluated on a separate sample set; thus, both a calibration development and a calibration evaluation sample set are required. Normally, a single group of samples is divided into two sub-sets and the development and evaluation performed at the one time; in reality, of course, the evaluation set continues indefinitely into the future.

4.1 Sample Selection

Decisions must be made at the outset as to the form in which samples are to be analysed; relevant variables could include the sample temperature, freshness, and whether ground or wholegrain material is to be used. If the former, a grinding procedure must be standardized (*e.g.* grinder type, sieve aperture size and feed rate) and the process of mixing and holding before NIR measurements are recorded must be specified rigidly. Sample pre-treatment, for example grain cleaning, must also be clearly described and adhered to.

Spectral acquisition in the NIR region is both non-destructive and rapid; on the contrary, generation of chemical data for calibration samples is slow and often expensive. For this reason, it is preferred to select those samples from the available set that contain the maximum amount of chemical and physical information about the material under study and thus minimize the expenditure on chemical analysis. A number of strategies have been advanced to facilitate this pre-selection process, all of which identify those spectra which are least similar to the others in the sample set, but it is not clear that such pre-selection techniques offer any significant advantages over operator or random selection processes. In any case, the set of samples used for calibration purposes should be describd as follows:

[8] W. L. Wang, L. R. Hudson and H. F. Tseng, *Opt. Eng.*, 1987, **26**, 844.
[9] D. C. Tilotta, R. M. Hammaker and W. G. Fately, *Appl. Spectrosc.*, 1987, **41**, 727.
[10] T. Hyvärinen, E. Herrala, J. Malinen and P. Niemalä, in 'Near Infra-red Spectroscopy. Bridging the Gap between Data Analysis and Applications', ed. K. I. Hildrum, T. Isaksson, T. Naes and A. Tandberg, Ellis Horwood, 1992, pp. 135–140.

(*a*) it should cover the widest possible variation in the constituent(s) of interest; for natural substances this does not usually cause a major problem but for processed materials, obtaining a range in composition may require unusual adjustments to the production process or the accumulation of out-of-specification samples over a prolonged time-period;

(*b*) the distribution of the constituent(s) of interest should not be normal but even across the range obtained; use of a normally distributed set of samples would result in a calibration model rich in information about those samples most commonly encountered but poor in experience of those at the extremes of the distribution which are arguably the most important;

(*c*) all normal sources of variation which may be encountered in future samples should be included; for example in the case of grain crops, such variables would include year of harvest, growing region, variety, disease conditions, weathering, presence of shrivelled grain *etc.* Any culling of the calibration sample set at this stage will result in a calibration of poor long-term predictive ability. As new sources of variation are encountered, for example the introduction of a new grain variety or a change in a production process, samples exhibiting this change should be collected and added to the calibration set so as to update the calibration and ensure its continued accuracy.

No fixing rules exist for the optimum number of samples required but typically 30–100 might be used for the prediction of a single constituent.

4.2 Sample Analysis

In the past, perhaps the greatest source of error in NIR calibrations lay in the performance of wet chemistry analyses. Since the spectra are regressed against the chemical data for model development (see below), the precision of the laboratory analysis is the practical lower limit to the prediction accuracy of the NIR calibration. Therefore, stringent quality control must be exercised over the chemical analyses through the use of blind duplicate (or even triplicate) testing, random sample testing order *etc.* Standard deviation between duplicate (SDD) laboratory measurements is a key statistic and must be continually monitored; it must also be reported together with the predictive descriptors of calibration performance (see below). It cannot be stressed too much that failure of NIR calibrations has been due more to unsatisfactory chemical analyses than to any other part of the calibration process. While marked improvements in laboratory performance have been seen over the last decade and are likely to continue due to the growing interest in laboratory accreditation, all laboratory results must be subject to scrutiny.

4.3 Instrument Performance Checks

Whether the NIR instrument being used for spectral collection is limited to a range of wavelengths (*e.g.* 19 or 20) or uses a scanning monochromator for full wavelength range recording, it is essential that instrument performance is checked before and after use and that a standard sample presentation technique is utilized throughout the spectral collection process.

For limited wavelength range instruments, good housekeeping practices are limited to ensuring the cleanliness of the sample drawer, ceramic tile standard and exit lens of the optic system. Regular air filter replacement, the location of the instrument in a stable atmosphere out of direct sunlight and away from draughts are all essential to ensure steady instrument performance. Short- and long-term instrument checks should be incorporated into the daily operating schedule.

Short-term instrument checks (precision testing) involve the regular determination of the standard deviation between duplicate readings (including re-packing) of a suitable material. Such determinations can be performed on samples analysed in the normal way during the working day or with a series of specific check samples. The precision thus determined should be stable over long time periods; poor precision figures (assuming correct sample mixing and presentation) may indicate instability in the mains power supply or in internal electronic components.

Long-term checks may be made both on real samples (once a calibration is available) or, depending on the instrument, using the internal ceramic disc. While the latter does not show significant absorption in the NIR region, any reading obtained from it with any given calibration should be stable over very long periods of time *e.g.* analysis of the ceramic disc fitted to Bran and Luebbe fixed-filter instruments using a wheat protein calibration should be stable to within $\pm 0.1\%$. Certain fixed-filter instruments allow recording of log $1/R$ (where R=reflectance) readings of the fitted ceramic tile to be stored in memory and compulsorily checked before each sample measurement. Significant deviation of these values from those stored generates an error message and prevents use of the instrument until the problem is rectified.

In the case of scanning instruments, more sophisticated diagnostic tests are supplied with the manufacturer's software; of particular importance in this regard may be mentioned instrument response (absolute reflectance of the ceramic disc), wavelength accuracy and repeatability (noise). Analysis of a check sample permanently packed in an air-tight cell is also recommended. Cumulative sum (CUSUM) techniques have advantages in this respect.[11] These instruments tend to be more sensitive to their environment than their less sophisticated counterparts and often the use of air-conditioned laboratories is recommended. The most recent generation of such equipment is, however, more robust due to the increased use of solid state electronics but a relatively steady environment and shielding from direct sunlight and draughts is nonetheless to be preferred. It is important to ensure that the instruments have attained normal working temperature before recording spectra; 1–2 h may be needed for this.

4.4 Sample Presentation

A characteristic of NIR spectroscopy has always been the wide range of sample presentation options available. While it is true that most work was initially

[11] G. Downey, *Lab. Pract.*, 1989, **38**(8). 72.

performed on ground material in reflectance mode, techniques involving transmission and transflectance are now available. A wide range of sample accessories can handle liquids, slurries, solids, films, wholegrains and whole fibrous material; together with the use of fibre optic accessories (single fibres or fibre bundles), remote reflectance, interactance immersion and surface interactance probes allow almost any type of sample to be conveniently and reproducibly handled.

Ground sample presentation is normally in a quartz-windowed sample cell. Variables affecting sample presentation in this manner include cell filling and packing density; instrument manufacturers have designed such cells so as to render sample packing density relatively constant as long as the cell is filled correctly. This procedure is not difficult and can normally be standardized for any operator in a short time but training and consistency must not be neglected. In the case of liquids, sample temperature and, for dairy products, homogenization must be critically controlled for reproducible spectral acquisition. In fact, sample temperature is a variable which significantly affects peak location in the NIR region, especially for water. In the past, the practice has been to standardize on sample temperature and only analyse samples which have been thus equilibrated; an alternative method which may better handle unknown samples is to vary the temperature of calibration samples so as to build sample temperature into the calibration. In this way, temperature variations in unknown samples tested in the future may be handled efficiently.

A novel mode of liquid presentation is DESIR – dry extract system for infrared.[12] In this method, a small volume of liquid extract is placed on a glass-fibre disc and the solvent evaporated using hot air. The disc plus dried extract is then measured in reflectance or transmittance mode after a short cooling period. A modification of this procedure for use with slurries and pastes has been published[13] – SPOT, solid particle overlay technique. Methods for microsampling of liquids[14] and the analysis of single pharmaceutical tablets[15] have been reported.

4.5 Spectral Collection

General rules, which must be adhered to for satisfactory spectral collection, involve the measurement of replicate (normally duplicate will suffice) sample scans (with sample repacking between scans) with the storage of mean spectra and, in the case of scanning instruments, examination of spectra in advance of any data manipulation. Samples should be analysed spectroscopically quite soon before or after they are analysed chemically to avoid changes in sample composition affecting the validity of the spectra; to facilitate this, spectral collection may need to be extended over several days given that chemical analysis is generally significantly slower than spectroscopy.

[12] M. Meurens and G. Alvaro, in 'Proceedings of the Third International Conference on Near Infrared Spectroscopy', Agricultural Publishing Centre, Gembloux, Belgium, 1990, vol. 1, p. 39–43.

[13] M. Meurens, in 'Near Infra-red Spectroscopy. Bridging the Gap between Data Analysis and Applications', ed. K. I. Hildrum, T. Isaksson, T. Naes and A. Tandberg, Ellis Horwood, 1992, pp. 159–164.

[14] R. A. Lodder and G. M. Hieftje, *Appl. Spectrosc.*, 1988, **42**, 518.

[15] R. A. Lodder and G. M. Hieftje, *Appl. Spectrosc.*, 1988, **42**, 556.

Spectral examination should involve visual inspection of acquired spectra and mathematical analysis. The former is rapid and allows for the detection of gross spectral anomalies that may arise from time to time while the latter can be used to confirm that the spectral collection is relatively homogeneous. An example of the latter involves calculation of the location of the mean spectrum of a given collection in multidimensional space and the Mahalanobis distance of each spectrum in the collection from the mean; any spectra lying more than three times the standardized H distance (Mahalanobis distance) from the mean are flagged as potential outliers and may be re-analysed to confirm whether they belong to this sample set.[16]

5 CALIBRATION DEVELOPMENT

While, in principle, spectra collected using limited wavelength and full range scanning instruments are subject to the same mathematical treatments for calibration development, in practice different approaches are used. These will be dealt with separately in this text.

5.1 Limited Wavelength Range Instruments

These instruments probably still account for most of the calibrations in use today. The number of wavelengths involved varies from 4–20 with calibrations normally using from 3–6/7 of these. In the 1960s, Karl Norris of the USDA demonstrated that quantitative predictions of moisture, protein and oil content in biological material could be made by means of stepwise multilinear regression (SMLR) techniques using absorbance values at a limited number of NIR wavelengths. This procedure first determines those wavelengths that are highly correlated with the constituent of interest and then compensates for the overlapping nature of NIR spectra by using readings from several of these wavelengths to estimate this constituent. A number of variants of the procedure exist, however, and the analyst has a rather bewildering choice between stepwise forward, stepwise backward, best single wavelength start, best double, triple *etc.* and finally the best all-possible-combination solution using two, three, four *etc.* wavelengths. Thus, even with the current generation of fast microprocessors, calibration development time can be extensive; a more serious problem, however, is that it is often not possible to select the 'best' calibration unambiguously. A large number of candidate calibrations is developed which must then be evaluated on a separate sample set to determine which is the most accurate and stable.

Light scatter, that most significant interference in NIR measurements, may only be compensated for to a limited degree by SMLR through inclusion in the regression equation of at least one reference wavelength, *i.e.* a wavelength at which the constituent of interest has no significant absorption. Measurements at either 1680 or 2310 nm are conventionally included for this purpose. SMLR has, in addition, a number of serious theoretical problems when dealing with NIR data;

[16] J. S. Shenk and M. O. Westerhaus, *Crop Sci.*, 1991, **31**, 469.

among these may be mentioned its inappropriateness when treating highly collinear data and the difficulty of detecting outlying (or unusual) samples.

The net effect of the former is to give rise to unstable calibrations – this may be overcome by constant updating of the calibration development set of samples so as to maximize its information content but the workload involved in this activity can be significant. As an example of this, all the calibrations produced in Ireland for the prediction of protein in native wheat[17] and malting barley[18] have been developed using grain samples taken from several harvests. As the calibration set grew after each year, the information content of this set became more complete and the prediction errors calculated for each new grain crop decreased and then stabilized. As regards the problem with outlier detection, unacceptable prediction errors may be wrongly ascribed to poor calibration performance rather than the analysis of a number of samples which are not typical of those in the calibration development set. Thus the entire calibration comes into disrepute rather than highlighting the fact that one or more samples possess unique spectral characteristics, which merit separate study.

Despite these problems, SMLR has been the method of choice for the vast majority of calibrations currently used in the agricultural and food processing industries. When used correctly, it generally produces satisfactory calibrations and it has the added advantage of being relatively-easily comprehended by non-mathematicians.

The mathematical (least squares) model used takes the form:

$$y = a + b_1x_1 + b_2x_2 + \;\ldots\ldots\; b_nx_n$$

where y is the concentration of the constituent to be predicted, a is a constant, b_1–b_n *etc.* are regression coefficients determined by the model and x_1–x_n *etc.* are intensity readings (normally log $1/R$ where R is reflectance) for the sample at wavelengths 1–*n*. Experience has shown that fewer rather than more wavelengths should be used to maximize the likelihood of calibration stability – typically 3–6/7 are favoured but the optimum for each application must be determined by trial and error.

Indicators of the goodness-of-fit of the fitted regression line are the residual standard deviation (often referred to as the standard error of calibration, SEC) and the correlation coefficient, *r*. The SEC is defined[1] as follows:

$$\mathrm{SEC} = \sqrt{\sum_{i=1}^{n} (y_\mathrm{i} - a - bx_\mathrm{i})^2 / n - 2}$$

where y_i and x_i are the predicted and actual values for sample i, *a* and *b* are constants and *n* is the number of samples involved. The figure obtained is only a guide to the performance of the calibration on unknown samples and is normally

[17] G. Downey and S. Byrne, Protein determination of wheat in trade by near infra-red reflectance : calibration and instrument performance over a 4-year period, *Sci. Aliments*, 1987, **7**, 325.

[18] G. Downey, Development, evaluation and collaborative testing of calibrations to predict protein and moisture in ground barley by near infra-red reflectance, *Ir. J. Food Sci. Technol.*, 1986, **10**(2), 119.

better than the accuracy which may be achieved on such material. The correlation coefficient is influenced by the range of concentration of the constituent of interest; the closer its value is to 1.0, the better the performance potential of the calibration. Its value has in the past been considerably over-emphasized however, and it does not indicate the predictive accuracy of a calibration.

5.2 Full Spectrum Methods

With the improved performance and availability of scanning NIR equipment, increasing numbers of applications are being made using full spectral scans, *i.e.* with up to 700 data points per sample. The use of SMLR with such data can be particularly problematic, especially if wavelengths selected for inclusion in a calibration model lie at the top of narrow absorbance peaks. Overfitting of calibrations (*i.e.* fitting incidental changes in spectral data to variations in chemical composition) also becomes a major potential problem.

One way of reducing this intercorrelation to separate overlapping absorbance bands and to correct for baseline variations is to use derivative spectra[2] rather than raw log $1/R$ data. In derivative spectra, plots of absorbance against wavelength are replaced by plots of the rate of change of absorbance against wavelength; calculation of these rates of change may be done by simply computing the difference between absorbances at equally spaced intervals, *e.g.* 10 nm gap. In practice, however, some amount of smoothing using a pre-determined segment size is simultaneously applied. Second derivative spectra are easier to interpret and find more application than first or higher orders, which become increasingly sensitive to spectral noise and wavelength instability. Discovery of the optimum gap and segment sizes used in derivative calculation must be done by trial and error and adds to the computer time necessary to develop calibrations. Nevertheless, the use of derivatives was a major step forward in generating stable calibrations by SMLR using full spectral scans for a significant period of time.

Considerable effort has been expended by a number of NIR researchers on the development of more satisfactory mathematical approaches to calibration development – this has coincided with the growth of a new field of applied statistics known as chemometrics.

Several global calibration methods have been developed to deal with these large data sets but a few have gained particular acceptance – these are principal components analysis (PCA) and partial least-squares in either of its two variants (PLS1 or 2). FT methods have been reported[19] but their application has been limited due to the rapid reduction in the cost of computer memory (they were one of the earliest forms of data compression applied to NIR spectra), enormous increase in microprocessor speed and, perhaps most importantly, the conceptual difficulties attached to their use.

Various mathematical approaches aimed at reducing the problem of light scatter have also been published.

5.2.1 Data Pre-treatment. The minimization of light scatter effects in full spectral scans has been approached by several workers as a data pre-treatment step.

[19] W. F. McClure, A. Hamid, F. G. Giesbrecht and W. W. Weeks, *Appl. Spectrosc.*, 1984, **38**, 322.

Among these may be mentioned double centring of the spectral data matrix,[20] multiplicative scatter correction[21] and standard normal variate transformation.[22] As is so often the case in NIR applications development, no hard and fast rule exists to recommend any one of these treatments over any other or indeed to confirm the advantage of using any of them as compared to the raw optical data. Given that a number of them are supplied with chemometric software packages, a sensible step is to try one or other and examine the effect in relation to the success of any model developed.

5.2.2 Principal Component Analysis (PCA). Principal component analysis (PCA) reduces intercorrelation between the wavelength data by reducing the 700 data points to a much smaller number (usually 20 or less) of synthetic variables. These synthetic variables are orthogonal (*i.e.* it is a condition of their calculation that the component axes are not intercorrelated) thus reducing the danger of over-fitting when the principal components are regressed against chemical data to facilitate prediction of composition. They are also linear combinations of the original variables and hence contain information from the entire spectrum.

Each principal component calculated describes the major spectral variation remaining in the dataset; thus the first principal component, which is in fact the mean of the dataset, accounts for most of the initial variance (generally $>90\%$), the second the next greatest quantity *etc.* Those components which are last to be fitted account mainly for non-systematic noise. The extent of the spectral variation accounted for by any individual component does not, however, relate in any simple way to the importance of this component in a particular regression model. For example, the variation explained by the first component is generally dominated by that due to particle size effects. Since the object is to fit a property (*e.g.* protein content) to the spectra, this first factor may have a relatively low correlation. A key feature of this data reduction technique is that it is performed on the spectra alone, *i.e.* results from chemical analysis play no part in the calculation of components.

Since each principal component is a linear combination of all the original variables, they may be plotted against wavelength to give a curve analogous to a spectrum. The major features of each component (see Figure 2) may thus be interpreted in terms of chemical structures within the material under investigation[23] although this is normally difficult for components beyond the third.

Principal components may be input to a multiple linear regression model to produce a prediction equation. Although the components are orthogonal, the danger of over-fitting in the regression does still exist and care must be exercised in the selection of the stopping rule for inclusion of new components. A convenient guide is when the introduction of a new term is not significant at the 5% level. As always, evaluation of the performance of a principal component regression (PCR) equation must be done using a separate prediction sample set.

Paradoxically, one disadvantage of PCR lies in the inclusion of all the spectral data; models thus developed may not be transferred to limited wavelength instruments.

[20] G. Downey, P. Robert, D. Bertrand and M-F. Devaux, *J. Sci. Food Agric.*, 1987, **41**, 219.
[21] P. Geladi, D. MacDougall and H. Martens, *Appl. Spectrosc.*, 1985, **39**(3), 491.
[22] R.J. Barnes, M.S. Dhanoa and S.J. Lister, *Appl. Spectrosc.*, 1989, **43**(5), 772.
[23] I.A. Cowe and J.W. McNicol, *Appl. Spectrosc.*, 1985, **39**, 257.

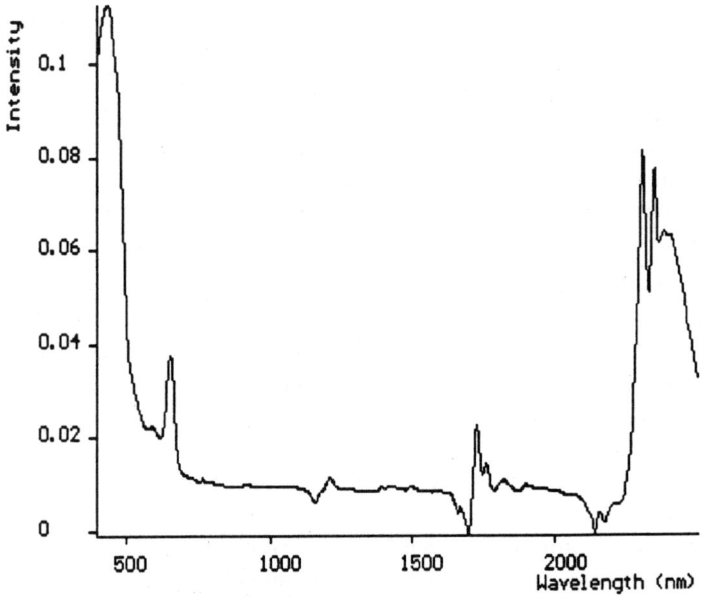

Figure 2 *Spectral pattern of principal component function after olive oil analysis*

5.2.3 Partial Least-squares (PLS). PLS is a recently developed data compression technique[24] and is being applied increasingly to NIR data. It is analogous to PCA in that a reduced number of factors is calculated to explain the variance in a dataset but while PCA does this independently of the chemical data, PLS involves both the spectral and chemical data at the same time. Thus, PLS primarily describes the variables in the NIR data, which are relevant for describing the variation in the chemical data. PLS1 is a procedure that allows for the investigation of one non-spectral variable at a time while the more recent PLS2 permits several constituents to be modelled at the same time.

One result of this involvement of the chemical data in factor calculation is that variation in the spectral set which is poorly or not at all correlated with chemical composition is relegated to the later PLS factors; in PCA, this relegation may not be so marked if this variation is significant within the spectral data.

PLS is finding increasing favour in NIR applications mainly because it is faster than PCA, which is an iterative process. In general, there is little to choose between the performance of the two.

5.2.4 Neural Networks. The most recent chemometric strategy to be applied to NIR data is that of neural networks. Few applications have been reported although one of specific interest to NIR users has appeared.[25] OMNIS (optimal minimal neural interpretation of spectra) uses principal components as inputs to a

[24] S. Wold, H. Martens and H. Wold, 'Matrix Pencils. Lecture Notes in Mathematics'. Springer Verlag, Heidelberg, 1982.

[25] C. Borggaard and H. H. Thodberg, *Anal. Chem.*, 1992, **62**(5), 545.

neural network, it has direct connections between input and output, the network size is optimized to ensure only the minimum number of connections necessary to describe the data is used and cross-validation is used systematically for network optimization.

Only a very limited assessment of the efficacy of neural networks may be made at this moment but their main advantage seems to lie in the way they handle non-linear data. Difficulties associated with the number of iterations used in model development seem no longer to be a significant problem due to the increasing power of personal computers.

5.2.5 Discriminant Methods. Qualitative analysis is an important goal for the food and other industries since a common analytical problem is the correct assignment of an unknown sample to one of a number of functionally distinct groups rather than the need for quantitative information. Historically however, the chemical and spectroscopic approach to this problem has been to use such quantitative information (*e.g.* protein content and degree of oxidation) as an empirical tool for assignment. While often useful, the challenge has been to bypass this intermediate step and perform the classification directly on the basis of the spectral properties of a sample.

A number of workers have developed statistical tools which aim to classify the spectrum of a given sample in multidimensional space in relation to known categories of relevant material by using multivariate techniques. The first report of such a procedure[26] used Mahalanobis distances to categorize chemical products (using a fixed-filter instrument) while another[27] utilized multiple discriminant analysis for the identification of wheat varieties. More recently, a procedure using factorial discriminant analysis has been applied to the classification of milk powders[28] while canonical variate analysis was used for discrimination of chocolate quality.[29]

This is an applications area which seems likely to grow in the future, especially given that it is now being used by the pharmaceutical[30] and chemical industry for product identification and that the FDA has recently approved the use of NIR analysis for the release testing of ampicillin trihydrate, a bulk pharmaceutical product for human consumption.[31] A useful review of this subject area has recently been published.[32]

6 CALIBRATION EVALUATION

Evaluation of all derived calibrations must be performed on an independent sample set. This should be representative of all such samples likely to be encountered in the future and thus, unlike the development sample set, should be

[26] H. L. Mark and D. Tunnell, *Anal. Chem.*, 1985, **57**, 1449.
[27] P. Dardenne and R. Biston, 'Attempt to Recognise Wheat Species by Discriminant Analysis' presented at the International NIR/T Spectroscopy Conference, Budapest, Hungary, 1986.
[28] G. Downey, P. Robert, D. Bertrand and P. Kelly, *Appl. spectrosc.*, 1990, **44**(1), 150.
[29] A. M. C. Davies, J. G. Franklin, A. Grant, N. M. Griffiths, R. Shepherd and G. R. Fenwick, *Vib. Spectrosc.*, 1991, **2**, 161.
[30] W. W. Ciurczak, *NIR News*, 1991, **2**(4), 8.
[31] Anon., *NIR News*, 1993, **4**(2), 7.
[32] G. Downey, *Analyst (Cambridge)*, 1994, **119**, 2369.

normally distributed with respect to the constituent of interest. It should also contain variations typical of such samples.

Statistical evaluation of calibrations is achieved by simple linear regression of the predicted *vs.* the reference chemical values for the evaluation sample set:

$$y = a + bx$$

where y is the chemical (reference) value for a given sample and x is the predicted (NIR) value. As with calibration development, strict quality control must be maintained over the chemical analyses, which should also be performed in duplicate or triplicate.

The number of statistical parameters used to evaluate calibration performance has varied from author to author but the following represent those of greatest importance.

The slope of the fitted regression line, b, should not be significantly different from 1.0; if it is, then the calibration is said to be skewed. Skewed calibrations can be corrected by adjustments to the intercept terms and the regression constants in the calibration model but the presence of a skew should be taken as an indication that the calibration model is incomplete and re-calibration is advised.

The intercept term, a, should not be significantly different from zero; if it is, the calibration is said to exhibit a bias. This may be simply corrected by subtraction or addition of a constant term to or from the predicted results and may or may not be adjudged to be serious. The presence of a significant bias term is common, for example, in the case of biological materials from different harvests and adjustment at the outset of a harvest often solves the problem. In a process product, a bias may suggest a change in, for example, some process parameter; while the presence of a bias does not invalidate a calibration, its cause should be elucidated and removed or incorporated.

If both a and b are satisfactory, the most meaningful statistic to be reported is the standard error of prediction (root mean square prediction error) which is defined as follows:

$$\text{SEP} = \sqrt{\left(\sum_{1}^{n} (y_{\text{obs}} - y_{\text{pred}})^2 / n\right)}$$

where n is the number of samples analysed.

Ideally, the standard error of prediction (SEP) will be close to the standard error of the calibration set and if produced correctly will give a realistic assessment of the performance of the calibration into the future.

Another indicator of performance is the correlation coefficient, r, between the reference and predicted results. For a perfect correlation, this will have a value equal to 1.0 but this is rarely, if ever, encountered. The higher the value the better, and the analyst can feel comfortable with values >0.9; if this value is low (<0.5) it is questionable whether NIR is a useful technique for prediction of the constituent being analysed.

The best calibration evaluation set is a set of samples collected over a given time period other than that during which the calibration development set were accumulated. For certain sample types this may not be possible and indeed perhaps only a limited number of samples may be collected in the first place. Calibration development and evaluation then may be performed using the technique of cross-validation, which operates as follows. In a set of 45 samples, sample #1 is removed and a calibration developed on the other 44; this calibration is then used to predict the composition of the deleted sample, #1. A difference between the reference and predicted result for this sample is obtained and recorded. This procedure is then repeated with the deletion of sample #2, the development of a calibration using samples 1 and 3–45, the prediction of sample #2 and calculation of the difference between predicted and reference data. The process is repeated for all 45 samples and an overall figure for the standard deviation of the differences between predicted and reference figures is calculated – this is analogous to the SEP described above.

Outlying samples are a perennial problem in NIR applications; they may be defined as those samples for which the residual (difference between reference and NIR result) is greater than three times the calibration SEP. Deleting such samples apparently improves the calibration performance but this step must only be taken when it can be justified by objective criteria. The correct initial approach must always be to repeat the analyses of these samples to check for measurement errors; if they still are poorly predicted and the operator is certain that they are actually samples which belong to the set of material being analysed, then they must remain in the set and should, in fact, be included in the calibration set since they obviously contain some information which that set is currently lacking.

7 STANDARDIZATION AND CONTROL

In most cases, calibrations are developed on a single NIR instrument with the aim of transferring them to a number of other instruments located in separate parts of a factory, region or even country. The question of calibration transferability was one which dogged early instrument generations and to which considerable effort has been devoted.

In the case of limited wavelength instruments, this is not a significant problem any longer, at least as regards transfer among instruments of the same model. In most cases, the only requirement is a slight bias adjustment; this observation has underpinned the successful commercial application of NIR for national grain purchasing schemes, for example.[33]

For scanning instruments, the problem is much more complex because of differences in both the wavelength and photometric scales in different spectrophotometers; calibration transfer between such instruments has only become a possibility very recently. The most recent generation of such equipment performs much more repeatably, thus reducing the size of the problem somewhat but in order to overcome what are essentially engineering shortcomings, the most

[33] G. Downey and S. Burne, *Ir. J. Food Sci. Technol.*, 1983, **7**, 135.

commonly used standardization approach has involved mathematical adjustment of the spectral data produced from one instrument ('slave') to duplicate that produced from another ('master') on the same sample set.[34] These adjustment parameters are stored in the software running the instrument and may be used or omitted when spectra are being recorded or manipulated. Another use of this cloning procedure is to allow the re-adjustment of the 'slave' instrument performance exactly to what it was prior to a major mechanical breakdown, for example.

Free-standing instruments require their performance to be monitored on a regular basis and a number of approaches may be used to this end. With earlier instrumentation, the most secure method involved the distribution of standard samples from a central laboratory and a comparison of the results obtained by each instrument with that held in the central location. When used in combination with daily checks on instrument performance (see above), this indirect method works quite effectively since, once the results for a given instrument are confirmed as accurate, this set of samples can be used as standards until the next batch are received sometime later. Such a method works well for the regulation of instruments throughout Ireland for analysis and purchase of the national grain crop (wheat and malting barley). With newer instrumentation, the facility for communicating electronically with a separate personal computer allows for the transfer of daily check sample results to the central control facility by telephone line; transfer of all test sample results together with courier delivery of a fixed percentage of these to the central laboratory for chemical analysis allows for tight control on a frequent basis.

Concomitant with this increased flow of information, however, comes the temptation to make frequent bias adjustments. These should never be made on the basis of less than ten samples nor for bias errors of less than 0.2% protein in the case of grain; over-zealous bias correction only introduces 'noise' into the system and becomes self-perpetuating.

8 THE FUTURE

Improvements in instrumentation inevitably mean a move from limited wavelength equipment to that which produces complete or near complete spectral scans. The concomitant increase in data set size presents a major mathematical problem for quantitative applications while qualitative problems require specific solutions. Chemometric approaches have permitted these problems to be addressed and have allowed the development of practical, useful models with a sound theoretical basis. Interpretative possibilities do exist through the examination of spectral patterns but this Achilles heel of NIR work remains to be firmly addressed. Now that protocols exist for the transfer of spectra between instruments and even between instruments operating in different spectral regions, it is becoming possible to compare spectra of the same sample taken in the mid-IR and NIR regions. Interpretation of such linked spectra are far from simple but out of the effort may

[34] J. S. Shenk and M. O. Westerhaus, *Crop. Sci.*, 1991, **31**, 1694.

finally emerge concrete ascriptions of NIR absorbances to defined chemical structures. Progress towards the production of miniature NIR instruments featuring innovative wavelength selection technologies and powerful, dedicated microprocessors may permit the move from NIR spectroscopy to NIR sensors with a consequent explosion in applications.

Subject Index